Simplified Design
of Structural Steel

‖‖‖‖‖‖‖‖‖‖‖‖‖‖‖‖‖‖‖‖‖‖‖‖‖‖‖‖‖‖‖‖‖‖‖‖

Simplified Design
of Structural Steel

IIIIIIIIIIIIIIIIIIIIIIIIIIIIIIIIIIIIIIIIIIIIIIIIIIIIIIIIIIIII

Harry Parker, M.S.

Emeritus Professor of Architectural Construction
University of Pennsylvania

FOURTH EDITION

prepared by

Harold D. Hauf, M.S.

Fellow American Institute of Architects
Fellow American Society of Civil Engineers

A Wiley-Interscience Publication

JOHN WILEY & SONS

New York   London   Sydney   Toronto

Library of Congress Cataloging in Publication Data:

Parker, Harry, 1887-
    Simplified design of structural steel.

    "A Wiley-Interscience publication."
    1. Building, Iron and steel. 2. Steel, Structural.
I. Hauf, Harold Dana. II. Title.

TA684.P33   1974   624'.1821   73-13562
ISBN 0-471-66432-4

Printed in the United States of America

10 9 8 7 6 5 4 3

Preface to the Fourth Edition

||||||||||||||||||||||||||||||||||||||||||||

Although design procedures for structural steel are becoming increasingly complex, there is continuing need for a less rigorous approach to the design of structural elements in buildings of moderate size, where many of the more sophisticated refinements permitted by modern codes are not warranted. Numerous individuals in offices of architects, designers, and builders, as well as those in training for such positions, require a working knowledge of structural steel design that will enable them to handle day-to-day structural problems that arise and that will provide background for working effectively with structural engineers on more complicated projects. The wide acceptance of the first three editions of *Simplified Design of Structural Steel* is cogent evidence of the validity of this point of view.

This fourth edition maintains the general spirit and approach declared by Professor Parker in the preface to the first edition, which is reprinted herewith. It has been written without employing advanced mathematics; a knowledge of high school algebra and arithmetic is adequate preparation for studying the design procedures presented. Basic principles of structural mechanics are reviewed as an aid to understanding the development of design procedures, and a major portion of the book is devoted to completely worked out illustrative examples. These are followed by similar problems to be solved by the reader.

This volume is used as a textbook in classrooms, and for home study. It is used in drafting rooms as a handy reference in the design of structures. Individuals preparing for civil service or state board examinations have found it useful as a refresher. With these applications in mind, particular care has been taken in the arrangement of the material, and answers have been provided for many of the exercise problems. The book contains several tables giving technical properties of structural shapes, safe loads, and other engineering data so that additional reference books are not required to understand the illustrative examples, nor to solve the exercise problems. However, the size and scope of the volume limit the number of such tables that can be included, so at some point during his study, the reader desiring more extensive data will find it advantageous to acquire a copy of the Seventh Edition of the *Manual of Steel Construction* published by the American Institute of Steel Construction (AISC).

The discussion and illustrative examples in this edition are keyed to the 1969 AISC *Specification for the Design, Fabrication & Erection of Structural Steel for Buildings.* Inasmuch as the preceding edition was based on the 1963 AISC Specification, and a new standard system for designations of structural shapes was adopted by the steel industry in 1970, recasting of much of the text has been necessary. Special acknowledgment is made to the American Institute of Steel Construction for granting permission to reproduce selected tabular data from the Seventh Edition of the *Manual of Steel Construction.*

The design procedures and engineering principles contained in this book follow the usual methods employed in practice based on elastic theory, commonly known as working stress design or allowable stress design. The brief chapter relating to plastic design is necessarily limited in scope; its purpose is to show the basis of the theory and to indicate how this concept can effect savings in the size of structural members. The AISC Specification is comprehensive and covers all aspects of both allowable stress design and plastic design. However, only a limited number of its provisions can be discussed in a book of this nature. The main purpose here is to explain methods used in determining adequate sizes for the common structural members that occur so frequently in the design of buildings and to provide a sound basis for more advanced study.

Phoenix, Arizona   *Harold D. Hauf*

Preface to the First Edition

||||||||||||||||||||||||||||||||||||||||||||||||

"Simplified Design of Structural Steel" is the fourth of a series of elementary books dealing with the design of structural members used in the construction of buildings. The first volume, "Simplified Engineering for Architects and Builders," discussed rather briefly the design of structural members of timber, steel, and reinforced concrete. The primary objective of the first volume was to present the basic principles of design for younger men having no preliminary training in engineering. The book being elementary in character and the scope limited, many important subjects were necessarily omitted.

The present volume treats of the design of the most common structural steel members that occur in building construction. The solution of many structural problems is difficult and involved but it is surprising, on investigation, how readily many of the seemingly difficult problems may be solved. The author has endeavored to show how the application of the basic principles of mechanics simplifies the problems and leads directly to a solution. Using tables and formulas blindly is a dangerous procedure; they can only be used with safety when there is a clear understanding of the underlying principles upon which the tables or formulas are based. This book deals principally in the practical application of engineering principles and formulas and in the design of structural members. The derivations of the most commonly used formulas are given in order that the reader may comprehend fully why certain formulas are appropriate in the solution of specific problems.

In preparing material for this book the author has assumed that the reader is unfamiliar with the subject. Consequently the discussions advance by easy stages, beginning with problems relating to simple direct stresses and continuing to the more involved examples. Most of the fundamental principles of mechanics are reviewed and, in general, the only preparation needed is a knowledge of arithmetic and high school algebra.

The text has been arranged for use in the classroom as well as for home study. The tables essential to structural engineering have been included so that no additional reference books are required.

In addition to discussions and explanations of design procedure, it has been found that the solution of practical examples adds greatly to the value of a book of this character. Consequently, a great portion of the text consists of the solution of illustrative examples. The examples are followed by problems to be solved by the student.

The author is deeply grateful to the American Institute of Steel Construction for its kindness in granting permission to reproduce tables and data from its manual "Steel Construction." In general, the American Institute of Steel Construction specifications have been followed in the preparation of this book. Thanks are extended also to the American Welding Society for permission to reproduce data from its "Code for Arc and Gas Welding in Building Construction," and to The Lincoln Electric Company of Cleveland, Ohio, publishers of "Procedure Handbook of Arc Welding Design and Practice." This company has graciously permitted the reproduction of the data and design procedure from its excellent and comprehensive volume. The cooperation of the societies and company mentioned above is greatly appreciated; without such cooperation a book of this character would not be possible.

The author proposes no new methods of design nor short cuts of questionable value. Instead, he has endeavored to present concise and clear explanations of the present-day design methods with the hope that the reader may obtain a foundation of sound principles of structural engineering.

Harry Parker

High Hollow, Southampton, Pa.
March 1945

Contents

||||||||||||||||||||||||||||||||||||||||||||||||||

Introduction

American Institute of Steel Construction	1
New Designations for Structural Shapes	2
Nomenclature	2
Abbreviations	5
Computations	6

1   Structural Shapes   7

1-1	General	7
1-2	Wide Flange Shapes	8
1-3	Standard I-Beams	9
1-4	Standard Channels	10
1-5	Angles	10
1-6	Structural Tees	11
1-7	Plates and Bars	12
1-8	Designations for Structural Shapes	13

2   Unit Stresses   25

2-1	General	25
2-2	Direct Stress	25
2-3	Kinds of Stress	27
2-4	Bending Stresses	29
2-5	Horizontal Shear	30

ix

2-6   Elastic Limit, Yield Point, and Ultimate Strength   32
2-7   Modulus of Elasticity   34
2-8   Factor of Safety   36
2-9   Allowable Stresses   37

3   Structural Steel   **39**

3-1   Grades of Structural Steel   39
3-2   Structural Carbon Steels   41
3-3   High-Strength Steels   41
3-4   Corrosion-Resistant Steels   42
3-5   Allowable Stresses for Structural Steel   42
3-6   Grades of Steel for Connectors   44

4   Reactions, Moments, and Shear   **45**

4-1   General   45
4-2   Types of Beam   45
4-3   Loading   47
4-4   Moments   49
4-5   Laws of Equilibrium   50
4-6   Determination of Reactions   53
4-7   Vertical Shear   57
4-8   Use of Shear Values   59
4-9   Shear Diagrams   59
4-10   Bending Moments   66
4-11   Bending Moment Diagrams   68
4-12   Positive and Negative Bending Moments   69
4-13   Concentrated Load at Center of Span   70
4-14   Simple Beam with Uniformly Distributed Load   71
4-15   Maximum Bending Moments   73
4-16   Overhanging Beams   76
4-17   Inflection Point   78
4-18   Cantilever Beams   81
4-19   Typical Loads for Simple and Cantilever Beams   82

5 Theory of Bending and Properties of Sections   85

5-1 Resisting Moment   85
5-2 The Flexure Formula   87
5-3 Properties of Sections   89
5-4 Centroids   89
5-5 Moment of Inertia   91
5-6 Section Modulus   92
5-7 Radius of Gyration   95
5-8 Properties of Built-Up Sections   95
5-9 Unsymmetrical Built-Up Sections   98

6 Use of the Beam Formula   101

6-1 Forms of the Equation   101
6-2 Investigation of Beams   102
6-3 Design of Beams for Bending   105
6-4 Structural Design Methods   107

7 Deflection of Beams   109

7-1 Deflection   109
7-2 Allowable Deflection   110
7-3 Deflection for Uniformly Distributed Loads   111
7-4 Concentrated Loads   114
7-5 A Convenient Deflection Formula   117
7-6 Deflection Found by Coefficients   118

8 Beam Design Procedures   121

8-1 General   121
8-2 Compact and Noncompact Sections   122
8-3 Lateral Support of Beams   123
8-4 Checking the Shear Stress   126
8-5 General Design Procedure   127
8-6 Examples Illustrating Beam Design   130
8-7 Safe Load Tables   139
8-8 Safe Load Table for Channels   146

8-9   Equivalent Tabular Loads   147
8-10   Laterally Unsupported Beams   150
8-11   Long Spans and Light Loads   152
8-12   Lintels   154
8-13   Separators   159
8-14   Beam Bearing Plates   159
8-15   Crippling of Beam Webs   165
8-16   Plate Girders   167

9   Floor Framing Systems   171

9-1   Layout   171
9-2   Dead Load   172
9-3   Live Load   174
9-4   Movable Partitions   174
9-5   Fireproofing for Beams   176
9-6   Design of Typical Framing   177
9-7   Open Web Steel Joists   182

10   Columns   189

10-1   Introduction   189
10-2   Column Shapes   190
10-3   Slenderness Ratio   191
10-4   Effective Column Length   192
10-5   Column Formulas   193
10-6   Allowable Column Loads   194
10-7   Design of Steel Columns   198
10-8   Steel Pipe Columns   199
10-9   Structural Tubing Columns   203
10-10 Double-Angle Struts   205
10-11 Eccentrically Loaded Columns   209
10-12 Bending Factors   210
10-13 Trial Section for Eccentrically Loaded Columns   210
10-14 Column Splices   212
10-15 Column Base Plates   213
10-16 Grillage Foundations   217

11   Bolted and Riveted Connections   **221**

11-1   General   221
11-2   Riveting   221
11-3   Failure of Riveted Joints   225
11-4   Shear Strength of Rivets   226
11-5   Bearing Strength of Rivets   226
11-6   Allowable Stresses for Rivets   230
11-7   Bolted Connections   230
11-8   Unfinished Bolts   231
11-9   High-Strength Bolts   232
11-10   Gage Lines   234
11-11   Pitch of Bolts and Rivets   235
11-12   Edge Distance   235
11-13   Net Sections   236
11-14   Framing Connection Details   238
11-15   Beam Connections to Girders   239
11-16   Framed Beam Connections   242
11-17   Beam Connections to Columns   245
11-18   Conventional and Moment Connections   247

12   Welded Connections   **249**

12-1   General   249
12-2   Electric Arc Welding   250
12-3   Welded Joints   250
12-4   Stresses in Welds   251
12-5   Design of Welded Joints   255
12-6   Beams with Continuous Action   258
12-7   Plug and Slot Welds   261
12-8   Miscellaneous Welded Connections   262
12-9   Symbols for Welds   265

13   Roof Trusses   **269**

13-1   General   269
13-2   Spacing of Trusses   272

13-3	Loads on Roof Trusses	272
13-4	Snow Load	274
13-5	Wind Load	275
13-6	Stress Diagrams	276
13-7	Wind Load Stress Diagram	279
13-8	Combinations of Loading	280
13-9	Equivalent Vertical Loading	280
13-10	Stresses Found by Coefficients	281
13-11	Design of a Roof Truss	283
13-12	Computation of Truss Loads	284
13-13	Purlin Design	285
13-14	Determination of Stresses	287
13-15	Identification of Truss Members and Joints	289
13-16	Design of Compression Members	289
13-17	Design of Tension Members	291
13-18	Rivets in Truss Members	291
13-19	Welded Truss Connections	294
13-20	End Bearing and Anchorage	297
14	Theory of Plastic Design	**299**
14-1	Elastic Design	299
14-2	Stress–Strain Diagram	300
14-3	Plastic Moment, Plastic Hinge	300
14-4	Plastic Section Modulus	303
14-5	Computation of Plastic Section Moduli	305
14-6	Shape Factor	306
14-7	Restrained Beams	309
14-8	Load Factor	311
14-9	Design of Simple Beams	312
14-10	Design of a Beam with Fixed Ends	313
14-11	Scope of Plastic Design	315
	Answers to Selected Problems	317
	Index	321

Simplified Design
of Structural Steel

||||||||||||||||||||||||||||||||||||||||

Introduction

IIIIIIIIIIIIIIIIIIIIIIIIIIIIIIIIIIIIIIIIIIIIII

American Institute of Steel Construction

The American Institute of Steel Construction (AISC) is the service organization for the fabricated structural steel industry in the United States. Due to the efforts of the AISC and other technical and professional organizations, most municipal building codes follow the provisions of the AISC *Specification for the Design, Fabrication & Erection of Structural Steel for Buildings*. Earlier editions of the AISC Specifications were widely adopted by building code authorities, and the 1969 revision is expected to follow the same pattern. Local codes should, of course, be consulted in actual design work since the frequency with which they are amended may control whether or not the new provisions have been incorporated.

The AISC *Manual of Steel Construction* is the basic handbook used by structural engineers. In addition to the AISC Specification and its accompanying Commentary, the Manual contains a vast amount of engineering data including more extensive tables than those presented in this book relating to dimensions, properties, and load-carrying capacities of rolled steel shapes. The Seventh Edition of the Manual is based on the 1969 AISC Specification,[1] and is a comprehensive presentation pertaining to new developments and

[1] The Sixth Edition was based on the 1963 Specification.

1

techniques of using the wide range of structural steels now available for building construction. The reader who desires to extend the scope of his study of steel design will find it essential to acquire a copy.

New Designations for Structural Shapes

In 1970, the Committee of Structural Steel Producers of the American Iron and Steel Institute promulgated a new standard system for designating structural shapes. The new designations are used throughout this book and in the Seventh Edition of the AISC Manual; they supersede the designations formerly used in practice and in the Sixth and previous Editions of the Manual. A listing of old and new designations for a few typical shapes is presented for comparison in Art. 1-8 of this volume.

Nomenclature

The general nomenclature of the AISC Specification differs somewhat from the symbols commonly used in structural mechanics. Although the latter are sometimes employed in this book in theoretical discussions, the former are used in design discussions. An abridgment of the AISC general nomenclature is given below for ready reference. In this listing, the units for stress values are indicated as kips per square inch (ksi) or pounds per square inch (psi), although the 1969 AISC Specification gives all values in ksi. The The complete general nomenclature is given in the AISC Manual.

A Cross-sectional area (sq in.)

       Area of beam or column base plate (sq in.)

A_f Area of compression flange (sq in.)

A_w Area of girder web (sq in.)

B_x, B_y Bending factor with respect to the X-X axis and Y-Y axis, respectively, for determining the equivalent axial load in columns subjected to combined loading conditions; equal to A/S_x and A/S_y, respectively

C_c Column slenderness ratio dividing elastic and inelastic buckling; equal to

$$\sqrt{\frac{2\pi^2 E}{F_y}}$$

E Modulus of elasticity of steel (29,000 kips per sq in.)

F_a Axial stress permitted in the absence of bending moment (ksi or psi)

F_{as} Axial compressive stress, permitted in the absence of bending moment, for bracing and other secondary members (ksi or psi)

F_b Bending stress permitted in the absence of axial force (ksi or psi)

F_p Allowable bearing stress (ksi or psi)
            Allowable bearing pressure on support (ksi or psi)

F_t Allowable tensile stress (ksi or psi)

F_v Allowable shear stress (ksi or psi)

F_{vw} Allowable shear stress in welds (ksi or psi)

F_y Specified minimum yield stress of the type of steel being used (ksi or psi). As used in AISC Specification, "yield stress" denotes either the specified minimum yield point (for those steels that have a yield point) or specified minimum yield strength (for those steels that do not have a yield point).

F_y' The theoretical maximum yield stress (ksi), based on the width-thickness ratio of one-half the unstiffened compression flange, beyond which a particular shape is not "compact." (AISC Specification Sect. 1.5.1.4.1, subparagraph b.)

F_y'' The theoretical maximum yield stress (ksi), based on the depth-thickness ratio of the web, beyond which a shape is not "compact." (AISC Specification Sect. 1.5.1.4.1, subparagraph d.) It is only applicable for cases of pure bending; i.e., $f_a = 0$.

F_y''' The theoretical maximum yield stress (ksi), based on the depth-thickness ratio of the web below which a particular shape may be considered "compact" for any condition of combined bending and axial stresses. (AISC Specification Sect. 1.5.4.1, subparagraph d.)

I Moment of inertia of a section (in.^{4})

K Effective length factor

L Span length (ft)

L_b Unbraced length of compression flange (ft)

L_c Maximum unbraced length of the compression flange at

	which the allowable bending stress may be taken at $0.66F_y$ (ft)
L_u	Maximum unbraced length of the compression flange at which the allowable bending stress may be taken at $0.60F_y$ (ft)
M	Moment (kip-ft or kip-in.)
M_D	Moment produced by dead load (kip-ft or kip-in.)
M_L	Moment produced by live load (kip-ft or kip-in.)
M_p	Plastic moment (kip-ft)
M_R	Beam resisting moment (kip-ft or kip-in.)
N	Length of bearing of applied load (in.)
N_e	Length at end bearing to develop maximum web shear (in.)
P	Applied load (kips)
P'	Equivalent axial load due to bending in members subject to axial compression and bending (kips)
R	Reaction (kips)
	Maximum end reaction for $3\frac{1}{2}$ in. of bearing (kips)
R_i	Increase in reaction (R) in kips for each additional inch of bearing
S	Elastic section modulus (in.³)
V	Statical (vertical) shear on beams (kips)
Z	Plastic section modulus (in.³)
b_f	Flange width of rolled beam or plate girder (in.)
c	Distance from neutral axis to extreme fiber of beams (in.)
d	Depth of beam or girder (in.)
f_a	Computed axial stress (ksi or psi)
f_b	Computed bending stress (ksi or psi)
f_p	Actual bearing pressure on support (ksi or psi)
f_t	Computed tensile stress (ksi or psi)
f_v	Computed shear stress (ksi or psi)
l	Actual unbraced length (in.)
l_b	Actual unbraced length in plane of bending (in.)
r	Governing radius of gyration (in.)
r_b	Radius of gyration about axis of concurrent bending (in.)
r_T	Radius of gyration of a section comprising the compression flange plus one-third of the compression web area, taken about an axis in the plane of the web (in.)
r_x	Radius of gyration with respect to the X-X axis (in.)

r_y	Radius of gyration with respect to the Y-Y axis (in.)
t	Girder, beam, or column web thickness (in.) .
t_f	Flange thickness (in.)
t_w	Web thickness (in.)
$\bar{x}$	Distance from the outside of the web to the minor (Y-Y) of a channel section (in.)
y	Distance from neutral axis to the outer-most fibers of cross section (in.)
	Distance from the back of the flange to the major (X-X) axis of a tee section (in.)
Δ	Beam deflection (in.)
kip	1000 pounds

Abbreviations

It will be observed that abbreviations are used in the preceding nomenclature to denote the units of measurement in which many of the items are expressed. Similar abbreviations are used throughout this book and are identified below for convenient reference.

Abbreviation	Quantity
cu ft	cubic foot
cu in.	cubic inch
ft	foot
ft-lb	foot-pound
in.	inch
in-lb	inch-pounds
kip	1000 pounds
kip-ft	kip-feet
kip-in.	kip-inches
ksf	kips per square foot
ksi	kips per square inch
lin ft	linear foot
lb	pounds
lb per cu ft	pounds per cubic foot
lb per lin ft	pounds per linear foot
psf	pounds per square foot
psi	pounds per square inch
sq ft	square foot
sq in.	square inch

The same abbreviation is used for both singular and plural, ft indicating either foot or feet. Also, it is common practice to omit the period generally used after abbreviations except in the case of inches (in.).

In many of the diagrams presented in this book, drawing symbols are used for convenience instead of letter abbreviations. Chief among these symbols are:

$$\# = \text{lb} \qquad '\# = \text{ft-lb} \qquad k' = \text{kip-ft}$$
$$\#/' = \text{lb per lin ft} \qquad ''\# = \text{in-lb} \qquad k'' = \text{kip-in.}$$

Computations

Slide-rule computations have been used throughout this book except in certain cases where extension of the numerical work seemed to clarify the explanation of an illustrative example. Three-significant-figure accuracy, attainable by use of the slide rule, gives results consistent with the assumptions of structural theory and conditions encountered in practice. If the reader does not possess a slide rule, he is urged to obtain one (with a pamphlet of instructions) at the first opportunity. The ability to use it is readily acquired, and in a short time it will become indispensable.

Confidence in the accuracy of one's computations is best gained by self-checking. However, in order to provide an occasional outside check, answers to some of the exercise problems are given at the end of the book. Such problems are indicated by an asterisk (*) following the problem number where it occurs in the text.

1
Structural
Shapes

||||||||||||||||||||||||||||||||||||||||||||||||||

1-1 General

The products of the steel rolling mills used as beams, columns, and other structural members are known as *sections* or *shapes*, and their designations are related to the profiles of their cross sections. American Standard I-beams (Fig. 1-1a) were the first beam sections rolled in the United States and are currently produced in sizes from 3 in. to 24 in. in depth. The wide flange shapes (Fig. 1-1c) are a modification of the I cross section and are characterized by parallel flange surfaces as contrasted with the tapered inside flange surfaces of Standard I-beams; they are available in depths from 4 in. to 36 in. In addition to the Standard I and wide flange sections, the structural steel shapes most commonly used in building construction are channels, angles, tees, plates, and bars. Tables 1-1 through 1-5 at the end of this chapter give the dimensions and weights of some of these shapes, together with other properties that are identified and discussed in Chapters 5 and 8. Complete tables of structural shapes are given in the AISC *Manual of Steel Construction*.

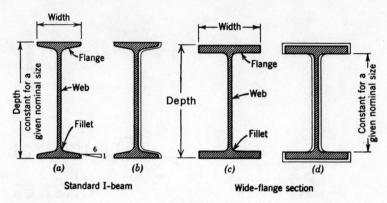

FIGURE 1-1

1-2 Wide Flange Shapes

In general, wide flange shapes have greater flange widths and relatively thinner webs than Standard I-beams; and as noted above, the inner faces of the flanges are parallel to the outer faces. These sections are identified in the new system of shape designations by the alphabetical symbol W, followed by the *nominal* depth in inches and the weight in pounds per linear foot. Thus, the designation W 12 × 27 indicates a wide flange shape of nominal 12-in. depth weighing 27 lb per lin ft. (In the old system its designation was 12 Wᶠ 27.)

The actual depths of wide flange shapes vary within the nominal depth groupings. By reference to Table 1-1, it is found that a W 12 × 27 has an actual depth of 11.96 in., while the depth of a W 12 × 36 is 12.24 in. This is a result of the rolling process during manufacture since the cross-sectional areas of wide flange shapes are increased by spreading the rolls both vertically and horizontally. The additional material is thereby added to the cross section by increasing flange and web thickness, as well as flange width (Fig. 1-1d). The resulting higher percentage of material in the flanges makes wide flange shapes more efficient structurally than Standard I-beams.[1] A wide

---

[1] The relative structural efficiency of different-shaped cross sections is discussed in more detail in Chapter 5.

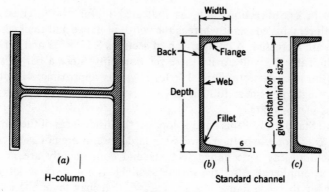

FIGURE 1-2

variety of weights is available within each nominal depth group.

In addition to shapes with profiles similar to the W 12 × 27, which has a flange width of 6.497 in., many wide flange shapes are rolled with flange widths approximately equal to their depths. The resulting H configurations of these cross sections (Fig. 1-2a) are much more suitable for use as columns than the I profiles. Referring to Table 1-1, it will be found that the following shapes, among others, fall into this category: W 14 × 87, W 12 × 65, W 10 × 60, and W 8 × 40. It is recommended that the reader compare these shapes with others listed in their respective nominal depth groups in order to visualize the variety of geometrical relationships.

1-3 Standard I-Beams

American Standard I-beams are identified by the alphabetical symbol S, the designation S 12 × 35 indicating a Standard shape 12 in. deep weighing 35 lb per lin ft. (In the old system, it was designated 12 I 35.) By referring to Table 1-2, it is found that this section has an *actual* depth of 12 in., a flange width of 5.078 in., and a cross-sectional area of 10.3 sq in. Unlike wide flange sections, Standard I-beams in a given depth group have uniform depths; and shapes of greater cross-sectional area are made by spreading the rolls in one direction only, as indicated in Fig. 1-1*b*. Thus, the depth remains constant whereas the width of flange and thickness of web are increased. A comparison of sections S 12 × 35 and S 12 × 50 in Table 1-2 will clarify these relationships. Since a bar of steel 1 sq in. in cross section and 1 ft long weighs approximately 3.4 lb, the weight per linear foot of any structural shape is 3.4 times the cross-sectional area.

All Standard I-beams have a slope on the inside faces of the flanges of 16⅔%, or 1 in 6. In general, Standard I-beams are not as efficient structurally as wide flange sections and consequently are not as widely employed. Also, the variety available is not nearly as large as that for the wide flange shapes. Factors that may favor the use of American Standard I-beams in any particular situation are their constant depths, narrow flanges, and relatively thicker webs.

1-4 Standard Channels

The profile of an American Standard channel is shown in Fig. 1-2*b*. These shapes are identified by the alphabetical symbol C. The designation C 10 × 20 indicates a Standard channel 10 in. deep weighing 20 lb per lin ft. Table 1-3 shows this section to have an area of 5.88 sq in., a flange width of 2.739 in., and a web thickness of 0.379 in. Like the Standard I-beams, the depth for a particular group remains constant, and the cross-sectional area is increased by spreading the rolls to increase flange width and web thickness as shown in Fig. 1-2*c*. Owing to their tendency to buckle when used independently as beams or columns, channels require lateral support or bracing. They are generally employed as elements of built-up sections such as columns and lintels;[2] however, the absence of a flange on one side makes channels particularly suitable for framing around floor openings.

1-5 Angles

Structural angles, as the name implies, are rolled sections in the shape of the letter L. Table 1-4 gives dimensions, weights, and other

---

[2] A lintel is a beam used over an opening in a masonry wall to support the load above. See Art. 8-12.

properties of equal leg angles, while Table 1-5 presents similar data for unequal leg angles. Both legs of an angle have the same thickness.

Angles are designated by the alphabetical symbol L, followed by the dimensions of the legs and their thickness. Thus, the designation L 4 × 4 × ½ indicates an equal leg angle with 4-in. legs ½ in. thick. Referring to Table 1-4, it is found that this section weighs 12.8 lb per lin ft and has an area of 3.75 sq in. Similarly, the designation L 5 × 3½ × ½ indicates an unequal leg angle with one 5-in. and on 3½-in. leg, both ½ in. thick. Table 1-5 shows that this angle weighs 13.6 lb per lin ft and has an area of 4 sq in. To change the weight and area for an angle of given leg length, the thickness of each leg is increased as shown in Fig. 1-3a. Thus, if the leg thickness of the L 5 × 3½ × ½ is increased to ⅝ in., Table 1-5 shows the resulting L 5 × 3½ × ⅝ to have a weight of 16.8 lb per lin ft and an area of 4.92 sq in. It should be noted that this method of spreading the rolls changes the leg lengths slightly.

Single angles are often used as lintels, and pairs of angles are widely employed as members of light steel trusses (Fig. 13-7). Angles were formerly used as elements of built-up sections such as plate girders (Fig. 8-16) and heavy columns (Fig. 10-2c), but the advent of the heavier wide flange shapes has largely eliminated their usefulness for this purpose. Short lengths of angles are commonly used as connecting members for beams and columns.

1-6   Structural Tees

A structural tee is made by splitting the web of a wide flange shape (Fig. 1-3c) or a Standard I-beam. The cut is normally made along the center of the web, producing tees with a stem depth equal to half the depth of the original section. Structural tees cut from wide flange

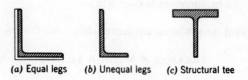

(a) Equal legs   (b) Unequal legs   (c) Structural tee

FIGURE 1-3

shapes are identified by the symbol WT, and those cut from Standard I shapes by ST. The designation WT 6 × 53 indicates a structural tee with a 6-in. depth weighing 53 lb per lin ft. This shape is produced by splitting a W 12 × 106 shape. Similarly, ST 9 × 35 designates a structural tee 9 in. deep, weighing 35 lb per lin ft, and cut from a S 18 × 70. No Tables of properties of these shapes are given in this book, but extensive tables are contained in the AISC Manual. Structural tees are used for the chord members of welded steel trusses (Fig. 13-8) and for the flanges in certain types of plate girders.

1-7   Plates and Bars

Plates and bars are made in many different sizes and are available in all of the structural steel specifications listed in Table 3-1.

Flat steel for structural use is generally classified as follows:

Bars: 6 in. or less in width, 0.203 in. and more in thickness
6 in. to 8 in. in width, 0.230 in. and more in thickness
Plates: Over 8 in. in width, 0.230 in. and more in thickness
Over 48 in. in width, 0.180 in. and more in thickness

Bars are available in varying widths and in virtually any required thickness and length. The usual practice is to specify bars in increments of $\frac{1}{4}$ in. for widths and $\frac{1}{8}$ in. in thickness.

For plates, the preferred increments for width and thickness are:

Widths: Vary by even inches, though smaller increments are obtainable.
Thickness· $\frac{1}{32}$ in. increments up to $\frac{1}{2}$ in.
$\frac{1}{16}$ in. increments over $\frac{1}{2}$ in. to 2 in.
$\frac{1}{8}$ in. increments over 2 in. to 6 in.
$\frac{1}{4}$ in. increments over 6 in.

The standard dimensional sequence when describing steel plate is:

$$\text{Thickness} \times \text{Width} \times \text{Length,}$$

with all dimensions given in inches, fractions of an inch, or decimals of an inch.

Column base plates and beam bearing plates may be obtained in the widths and thicknesses noted. For the design of column base plates and beam bearing plates, see Arts. 10-15 and 8-14, respectively.

1-8 Designations for Structural Shapes

As noted in Arts. 1-2 and 1-3, wide flange shapes are identified by the symbol W, and American Standard beam shapes by S. It was also pointed out that W shapes have essentially parallel flange surfaces while S shapes have a slope of approximately $16\frac{2}{3}\%$ on the inner flange faces. A third designation, M shapes, covers miscellaneous shapes that cannot be classified as either W or S; these have various slopes on their inner flange surfaces, and many of them are of only limited availability. Similarly, there are some rolled channels that cannot be classified as C shapes. These are designated by the symbol MC.

The list below gives designations for certain structural shapes, both in the new (1970) system and the former system. The latter are for reference only since they are now obsolete.

Type of shape	New designation	Old designation
Wide flange shapes	W 12 × 27	12 W 27
American Standard beams	S 12 × 35	12 I 35
Miscellaneous shapes	M 8 × 18.5	8 M 18.5
American Standard channels	C 10 × 20	10 [20
Miscellaneous channels	MC 12 × 45	12 × 4 [45.0
Angles—equal legs	L 4 × 4 × ½	∟ 4 × 4 × ½
Angles—unequal legs	L 5 × 3½ × ½	∟ 5 × 3½ × ½
Structural tees—cut from wide flange shapes	WT 6 × 53	ST 6 W 53
Structural tees—cut from American Standard beams	ST 9 × 35	ST 9 I 35
Structural tees—cut from miscellaneous shapes	MT 4 × 9.25	ST 4 M 9.25
Plate	PL ½ × 12	PL 12 × ½
Structural tubing: square	TS 4 × 4 × 0.375	Tube 4 × 4 × 0.375
Pipe	Pipe 4 Std.	Pipe 4 Std.

TABLE 1-1. Selected Wide Flange Sections*

W SHAPES

Properties for designing

Designation	Area A	Depth d	Flange		Web thickness t_w	Elastic properties					
						Axis X-X			Axis Y-Y		
			Width b_f	Thickness t_f		I	S	r	I	S	r
	In.^{2}	In.	In.	In.	In.	In.^{4}	In.^{3}	In.	In.^{4}	In.^{3}	In.
W 36 × 300	88.3	36.72	16.655	1.680	0.945	20300	1110	15.2	1300	156	3.8
× 135	39.8	35.55	11.945	0.794	0.598	7820	440	14.0	226	37.9	2.3
W 33 × 240	70.6	33.50	15.865	1.400	0.830	13600	813	13.9	933	118	3.64
× 118	34.8	32.86	11.484	0.738	0.554	5900	359	13.0	187	32.5	2.32
W 30 × 132	38.9	30.30	10.551	1.000	0.615	5760	380	12.2	196	37.2	2.2
× 99	29.1	29.64	10.458	0.670	0.522	4000	270	11.7	128	24.5	2.10
W 27 × 114	33.6	27.28	10.070	0.932	0.570	4090	300	11.0	159	31.6	2.18
× 84	24.8	26.69	9.963	0.636	0.463	2830	212	10.7	105	21.1	2.06
W 24 × 120	35.4	24.31	12.088	0.930	0.556	3650	300	10.2	274	45.4	2.78
× 110	32.5	24.16	12.042	0.855	0.510	3330	276	10.1	249	41.4	2.77
× 100	29.5	24.00	12.000	0.775	0.468	3000	250	10.1	223	37.2	2.75
× 76	22.4	23.91	8.985	0.682	0.440	2100	176	9.69	82.6	18.4	1.92
× 68	20.0	23.71	8.961	0.582	0.416	1820	153	9.53	70.0	15.6	1.87
W 21 × 73	21.5	21.24	8.295	0.740	0.455	1600	151	8.64	70.6	17.0	1.81
× 68	20.0	21.13	8.270	0.685	0.430	1480	140	8.60	64.7	15.7	1.80
× 62	18.3	20.99	8.240	0.615	0.400	1330	127	8.54	57.5	13.9	1.77
× 55	16.2	20.80	8.215	0.522	0.375	1140	110	8.40	48.3	11.8	1.73
× 49	14.4	20.82	6.520	0.532	0.368	971	93.3	8.21	24.7	7.57	1.31
× 44	13.0	20.66	6.500	0.451	0.348	843	81.6	8.07	20.7	6.38	1.27
W 18 × 85	25.0	18.32	8.838	0.911	0.526	1440	157	7.57	105	23.8	2.05
× 77	22.7	18.16	8.787	0.831	0.475	1290	142	7.54	94.1	21.4	3.04
× 55	16.2	18.12	7.532	0.630	0.390	891	98.4	7.42	45.0	11.9	1.67
× 50	14.7	18.00	7.500	0.570	0.358	802	89.1	7.38	40.2	10.7	1.65
× 45	13.2	17.86	7.477	0.499	0.335	706	79.0	7.30	34.8	9.32	1.62
W 16 × 50	14.7	16.25	7.073	0.628	0.380	657	80.8	6.68	37.1	10.5	1.59
× 45	13.3	16.12	7.039	0.563	0.346	584	72.5	6.64	32.8	9.32	1.57
× 40	11.8	16.00	7.000	0.503	0.307	517	64.6	6.62	28.8	8.23	1.56
× 36	10.6	15.85	6.992	0.428	0.299	447	56.5	6.50	24.4	6.99	1.52

* Compiled from data in the 7th Edition of the *Manual of Steel Construction*. Courtesy American Institute of Steel construction.

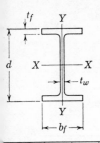

W SHAPES
Properties for designing

Nominal weight per ft.	r_T	$\dfrac{d}{A_f}$	Compact section criteria					Torsional constant J	Warping constant C_W	Plastic modulus	
			$\dfrac{b_f}{2t_f}$	F'_y	$\dfrac{d}{t_w}$	F''_y	F'''_y			Z_x	Z_y
Lb	In.			Ksi		Ksi	Ksi	In.⁴	In.⁶	In.³	In.³
300	4.46	1.31	4.96	—	38.9	—	43.7	64.2	398000	1260	241
135	2.97	3.75	7.52	48.2	59.4	48.0	18.7	7.03	68300	510	59.9
240	4.23	1.51	5.67	—	40.4	—	40.5	36.6	240000	919	182
118	2.87	3.88	7.78	45.0	59.3	48.2	18.8	5.32	48200	415	51.3
132	2.72	2.87	5.28	—	49.3	—	27.2	9.72	42100	437	58.5
99	2.61	4.23	7.80	44.7	56.8	52.6	20.5	3.78	26900	313	38.7
114	2.62	2.91	5.40	—	47.9	—	28.8	7.36	27600	343	49.4
84	2.52	4.21	7.83	44.4	57.6	51.1	19.9	2.79	17800	244	33.0
120	3.22	2.16	6.50	64.5	43.7	—	34.5	8.27	37500	338	69.9
110	3.20	2.35	7.04	54.9	47.4	—	29.4	6.45	33800	309	63.6
100	3.18	2.58	7.74	45.5	51.3	64.5	25.1	4.87	30100	280	57.2
76	2.32	3.90	6.59	62.8	54.3	57.5	22.4	2.70	11100	201	28.7
68	2.28	4.55	7.70	46.0	57.0	52.3	20.3	1.86	9350	176	24.4
73	2.16	3.46	5.60	—	46.7	—	30.3	3.02	7410	172	26.6
68	2.15	3.73	6.04	—	49.1	—	27.4	2.45	6760	160	24.4
62	2.13	4.14	6.70	60.7	52.5	61.6	24.0	1.83	5960	144	21.7
55	2.10	4.85	7.87	44.0	55.5	55.2	21.5	1.24	4970	126	18.4
49	1.63	6.00	6.13	—	56.6	53.0	20.6	1.09	2540	108	10.2
44	1.59	7.05	7.21	52.5	59.4	48.2	18.7	0.768	2120	95.3	10.2
85	2.37	2.28	4.85	—	34.8	—	54.4	5.50	7960	178	36.8
77	2.36	2.49	5.29	—	38.2	—	45.2	4.16	7070	161	33.1
55	1.98	3.82	5.98	—	46.5	—	30.6	1.66	3440	112	18.6
50	1.96	4.21	6.58	63.0	50.3	—	26.1	1.25	3050	101	16.6
45	1.94	4.79	7.49	48.5	53.3	59.7	23.2	0.889	2620	89.7	14.5
50	1.87	3.66	5.63	—	42.8	—	36.1	1.51	2260	91.8	16.3
45	1.85	4.07	6.25	—	46.6	—	30.4	1.11	1980	82.1	14.4
40	1.84	4.54	6.96	56.3	52.1	62.5	24.3	0.790	1730	72.8	12.7
36	1.81	5.30	8.17	40.8	53.0	60.4	23.5	0.545	1450	64.0	10.8

TABLE 1-1. Selected Wide Flange Sections (continued)
W SHAPES
Properties for Designing

Designation	Area A	Depth d	Flange Width b_f	Flange Thickness t_f	Web thickness t_w	Axis X-X I	Axis X-X S	Axis X-X r	Axis Y-Y I	Axis Y-Y S	Axis Y-Y r
	In.²	In.	In.	In.	In.	In.⁴	In.³	In.	In.⁴	In.³	In.
W 14 × 87	25.6	14.00	14.500	0.688	0.420	967	138	6.15	350	48.2	3.70
× 74	21.8	14.19	10.072	0.783	0.450	797	112	6.05	133	26.5	2.48
× 68	20.0	14.06	10.040	0.718	0.418	724	103	6.02	121	24.1	2.46
× 53	15.6	13.94	8.062	0.658	0.370	542	77.8	5.90	57.5	14.3	1.92
× 48	14.1	13.81	8.031	0.593	0.339	485	70.2	5.86	51.3	12.8	1.91
× 43	12.6	13.68	8.000	0.528	0.308	429	62.7	5.82	45.1	11.3	1.89
× 38	11.2	14.12	6.776	0.513	0.313	386	54.7	5.88	26.6	7.86	1.54
× 34	10.0	14.00	6.750	0.453	0.287	340	48.6	5.83	23.3	6.89	1.52
× 30	8.83	13.86	6.733	0.383	0.270	290	41.9	5.74	19.5	5.80	1.49
W 12 × 65	19.1	12.12	12.000	0.606	0.390	533	88.0	5.28	175	29.1	3.02
× 58	17.1	12.19	10.014	0.641	0.359	476	78.1	5.28	107	21.4	2.51
× 53	15.6	12.06	10.000	0.576	0.345	426	70.7	5.23	96.1	19.2	2.48
× 50	14.7	12.19	8.077	0.641	0.371	395	64.7	5.18	56.4	14.0	1.96
× 45	13.2	12.06	8.042	0.576	0.336	351	58.2	5.15	50.0	12.4	1.94
× 40	11.8	11.94	8.000	0.516	0.294	310	51.9	5.13	44.1	11.0	1.94
× 36	10.6	12.24	6.565	0.540	0.305	281	46.0	5.15	25.5	7.77	1.55
× 31	9.13	12.09	6.525	0.465	0.265	239	39.5	5.12	21.6	6.61	1.54
× 27	7.95	11.96	6.497	0.400	0.237	204	34.2	5.07	18.3	5.63	1.52
W 10 × 89	26.2	10.88	10.275	0.998	0.615	542	99.7	4.55	181	35.2	2.63
× 60	17.7	10.25	10.075	0.683	0.415	344	67.1	4.41	116	23.1	2.57
× 49	14.4	10.00	10.000	0.558	0.340	273	54.6	4.35	93.0	18.6	2.54
× 45	13.2	10.12	8.022	0.618	0.350	249	49.1	4.33	53.2	13.3	2.00
× 39	11.5	9.94	7.990	0.528	0.318	210	42.2	4.27	44.9	11.2	1.98
× 33	9.71	9.75	7.964	0.433	0.292	171	35.0	4.20	36.5	9.16	1.94
× 25	7.36	10.08	5.762	0.430	0.252	133	26.5	4.26	13.7	4.76	1.37
× 21	6.20	9.90	5.750	0.340	0.240	107	21.5	4.15	10.8	3.75	1.32
W 8 × 67	19.7	9.00	8.287	0.933	0.575	272	60.4	3.71	88.6	21.4	2.12
× 40	11.8	8.25	8.077	0.558	0.365	146	35.5	3.53	49.0	12.1	2.04
× 31	9.12	8.00	8.000	0.433	0.288	110	27.4	3.47	37.0	9.24	2.01
× 28	8.23	8.06	6.540	0.463	0.285	97.8	24.3	3.45	21.6	6.61	1.62
× 24	7.06	7.93	6.500	0.398	0.245	82.5	20.8	3.42	18.2	5.61	1.61
× 20	5.89	8.14	5.268	0.378	0.248	69.4	17.0	3.43	9.22	3.50	1.25
× 17	5.01	8.00	5.250	0.308	0.230	56.6	14.1	3.36	7.44	2.83	1.22

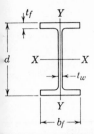

W SHAPES
Properties for Designing

Nominal weight per ft.	r_T	$\dfrac{d}{A_f}$	Compact section criteria					Torsional constant J	Warping constant C_W	Plastic modulus	
			$\dfrac{b_f}{2t_f}$	F_y'	$\dfrac{d}{t_w}$	F_y''	F_y'''			Z_x	Z_y
Lb.	In.			Ksi		Ksi	Ksi	In.⁴	In.⁶	In.³	In.³
87	4.02	1.40	10.5	24.5	33.3	—	59.4	3.68	15500	151	73.0
74	2.76	1.80	6.43	—	31.5	—	—	3.86	6000	126	40.5
68	2.74	1.95	6.99	55.7	33.6	—	58.4	3.01	5390	115	36.8
53	2.18	2.63	6.13	—	37.7	—	46.5	1.93	2540	87.1	21.9
48	2.16	2.90	6.77	59.4	40.7	—	39.8	1.44	2240	78.4	19.6
43	2.14	3.24	7.58	47.5	44.4	—	33.5	1.05	1950	69.7	17.3
38	1.80	4.06	6.60	62.5	45.1	—	32.5	0.796	1230	61.6	12.1
34	1.78	4.58	7.45	49.1	48.8	—	27.8	0.567	1070	54.6	10.6
30	1.75	5.37	8.79	35.3	51.3	64.4	25.1	0.376	886	47.2	8.95
65	3.31	1.67	9.90	27.8	31.1	—	—	2.19	5790	97.0	44.1
58	2.75	1.90	7.81	44.7	34.0	—	57.3	2.10	3580	86.5	32.6
53	2.74	2.09	8.68	36.2	35.0	—	54.1	1.59	3170	78.1	29.2
50	2.19	2.35	6.30	—	32.9	—	61.2	1.79	1880	72.5	21.4
45	2.18	2.60	6.98	55.9	35.9	—	51.3	1.32	1650	64.8	19.0
40	2.16	2.89	7.75	45.3	40.6	—	40.0	0.956	1440	57.5	16.8
36	1.77	3.45	6.08	—	40.1	—	41.0	0.830	873	51.6	11.9
31	1.75	3.98	7.02	55.4	45.6	—	31.7	0.536	728	44.1	10.1
27	1.74	4.60	8.12	41.3	50.5	—	25.9	0.351	611	38.0	8.62
89	2.88	1.06	5.15	—	17.7	—	—	7.74	4410	114	53.6
60	2.80	1.49	7.38	50.1	24.7	—	—	2.49	2670	75.0	35.1
49	2.77	1.79	8.96	33.9	29.4	—	—	1.38	2070	60.3	28.2
45	2.21	2.04	6.49	64.7	28.9	—	—	1.50	1200	54.9	20.2
39	2.19	2.36	7.57	47.6	31.3	—	—	0.971	995	46.9	17.1
33	2.16	2.83	9.20	32.2	33.4	—	59.2	0.580	792	38.8	14.0
25	1.56	4.07	6.70	60.7	40.0	—	41.3	0.373	320	29.6	7.30
21	1.53	5.06	8.46	38.1	41.3	—	38.8	0.210	246	24.1	5.77
67	2.33	1.16	4.44	—	15.7	—	—	5.05	1440	70.2	32.7
40	2.24	1.83	7.24	52.0	22.6	—	—	1.12	725	39.8	18.5
31	2.21	2.31	9.24	31.9	27.8	—	—	0.534	529	30.4	14.0
28	1.80	2.66	7.06	54.6	28.3	—	—	0.533	312	27.1	10.1
24	1.78	3.07	8.17	40.9	32.4	—	63.0	0.343	259	23.1	8.54
20	1.42	4.09	6.97	56.1	32.8	—	61.3	0.245	139	19.1	5.37
17	1.40	4.95	8.52	37.5	34.8	—	54.6	0.147	110	15.9	4.36

TABLE 1-2. American Standard I-Beams*
S. SHAPES
Properties for Designing

Designation	Area A	Depth d	Flange Width b_f	Flange Thickness t_f	Web thickness t_w	Axis X-X I	S	r	Axis Y-Y I	S	r
	In.²	In.	In.	In.	In.	In.⁴	In.³	In.	In.⁴	In.³	In.
S 24×120	35.3	24.00	8.048	1.102	0.798	3030	252	9.26	84.2	20.9	1.54
×105.9	31.1	24.00	7.875	1.102	0.625	2830	236	9.53	78.2	19.8	1.58
S 24×100	29.4	24.00	7.247	0.871	0.747	2390	199	9.01	47.8	13.2	1.27
×90	26.5	24.00	7.124	0.871	0.624	2250	187	9.22	44.9	12.6	1.30
×79.9	23.5	24.00	7.001	0.871	0.501	2110	175	9.47	42.3	12.1	1.34
S 20×95	27.9	20.00	7.200	0.916	0.800	1610	161	7.60	49.7	13.8	1.33
×85	25.0	20.00	7.053	0.916	0.653	1520	152	7.79	46.2	13.1	1.36
S 20×75	22.1	20.00	6.391	0.789	0.641	1280	128	7.60	29.6	9.28	1.16
×65.4	19.2	20.00	6.250	0.789	0.500	1180	118	7.84	27.4	8.77	1.19
S 18×70	20.6	18.00	6.251	0.691	0.711	926	103	6.71	24.1	7.72	1.08
×54.7	16.1	18.00	6.001	0.691	0.461	804	89.4	7.07	20.8	6.94	1.14
S 15×50	14.7	15.00	5.640	0.622	0.550	486	64.8	5.75	15.7	5.57	1.03
×42.9	12.6	15.00	5.501	0.622	0.411	447	59.6	5.95	14.4	5.23	1.07
S 12×50	14.7	12.00	5.477	0.659	0.687	305	50.8	4.55	15.7	5.74	1.03
×40.8	12.0	12.00	5.252	0.659	0.472	272	45.4	4.77	13.6	5.16	1.06
S 12×35	10.3	12.00	5.078	0.544	0.428	229	38.2	4.72	9.87	3.89	0.980
×31.8	9.35	12.00	5.000	0.544	0.350	218	36.4	4.83	9.36	3.74	1.00
S 10×35	10.3	10.00	4.944	0.491	0.594	147	29.4	3.78	8.36	3.38	0.901
×25.4	7.46	10.00	4.661	0.491	0.311	124	24.7	4.07	6.79	2.91	0.954
S 8×23	6.77	8.00	4.171	0.425	0.441	64.9	16.2	3.10	4.31	2.07	0.798
×18.4	5.41	8.00	4.001	0.425	0.271	57.6	14.4	3.26	3.73	1.86	0.831
S 7×20	5.88	7.00	3.860	0.392	0.450	42.4	12.1	2.69	3.17	1.64	0.734
×15.3	4.50	7.00	3.662	0.392	0.252	36.7	10.5	2.86	2.64	1.44	0.766
S 6×17.25	5.07	6.00	3.565	0.359	0.465	26.3	8.77	2.28	2.31	1.30	0.675
×12.5	3.67	6.00	3.332	0.359	0.232	22.1	7.37	2.45	1.82	1.09	0.705
S 5×14.75	4.34	5.00	3.284	0.326	0.494	15.2	6.09	1.87	1.67	1.01	0.620
×10	2.94	5.00	3.004	0.326	0.214	12.3	4.92	2.05	1.22	0.809	0.643
S 4×9.5	2.79	4.00	2.796	0.293	0.326	6.79	3.39	1.56	0.903	0.646	0.569
×7.7	2.26	4.00	2.663	0.293	0.193	6.08	3.04	1.64	0.764	0.574	0.581
S 3×7.5	2.21	3.00	2.509	0.260	0.349	2.93	1.95	1.15	0.586	0.468	0.516
×5.7	1.67	3.00	2.330	0.260	0.170	2.52	1.68	1.23	0.455	0.390	0.522

* Taken from the 7th Edition of the *Manual of Steel Construction.* Courtesy American Institute of Steel Construction.

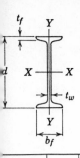

S SHAPES
Properties for Designing

Nominal weight per ft.	r_T	$\dfrac{d}{A_f}$	Compact section criteria					Torsional constant J	Warping constant C_W	Plastic modulus	
			$\dfrac{b_f}{2t_f}$	F'_y	$\dfrac{d}{t_w}$	F''_y	F'''_y			Z_x	Z_y
Lb.	In.	In.		Ksi		Ksi	Ksi	In.⁴	In.⁶	In.³	In.³
120	1.93	2.71	3.65	—	30.1	—	—	13.0	11000	299	36.4
105.9	1.93	2.77	3.57	—	38.4	—	44.8	10.4	10200	274	33.5
100	1.65	3.80	4.16	—	32.1	—	64.0	7.63	6390	240	24.0
90	1.65	3.87	4.09	—	38.5	—	44.6	6.05	6010	222	22.3
79.9	1.66	3.94	4.02	—	47.9	—	28.8	4.90	5660	205	20.7
95	1.70	3.03	3.93	—	25.0	—	—	8.46	4520	194	24.7
85	1.69	3.09	3.85	—	30.6	—	—	6.63	4200	179	22.8
75	1.46	3.96	4.05	—	31.2	—	—	4.60	2730	153	16.6
65.4	1.45	4.05	3.96	—	40.0	—	41.3	3.50	2530	138	15.2
70	1.41	4.17	4.52	—	25.3	—	—	4.15	1810	125	14.4
54.7	1.41	4.34	4.34	—	39.0	—	43.3	2.37	1560	105	12.1
50	1.31	4.28	4.53	—	27.3	—	—	2.12	811	77.1	9.97
42.9	1.31	4.38	4.42	—	36.5	—	49.6	1.54	743	69.3	9.02
50	1.31	3.32	4.15	—	17.5	—	—	2.82	404	61.2	10.3
40.8	1.25	3.46	3.98	—	26.0	—	—	1.76	436	53.1	8.85
35	1.20	4.34	4.67	—	28.0	—	—	1.08	324	44.8	6.79
31.8	1.20	4.41	4.60	—	34.3	—	56.2	0.901	307	42.0	6.40
35	1.15	4.12	5.03	—	16.8	—	—	1.29	189	35.4	6.22
25.4	1.13	4.37	4.74	—	32.2	—	63.9	0.604	153	28.4	4.96
23	0.987	4.51	4.90	—	18.1	—	—	0.551	61.8	19.3	3.68
18.4	0.973	4.70	4.70	—	29.5	—	—	0.336	53.5	16.5	3.16
20	0.914	4.63	4.92	—	15.6	—	—	0.451	34.6	14.5	2.96
15.3	0.894	4.88	4.67	—	27.8	—	—	0.241	28.8	12.1	2.44
17.25	0.845	4.69	4.97	—	12.9	—	—	0.374	18.4	10.6	2.36
12.5	0.817	5.02	4.64	—	25.9	—	—	0.168	14.5	8.47	1.85
14.75	0.761	4.66	5.03	—	10.1	—	—	0.323	9.09	7.42	1.88
10	0.741	5.10	4.60	—	23.4	—	—	0.114	6.64	5.67	1.37
9.5	0.684	4.88	4.77	—	12.3	—	—	0.120	3.10	4.04	1.13
7.7	0.662	5.13	4.54	—	20.7	—	—	0.073	2.62	3.51	0.964
7.5	0.621	4.60	4.83	—	8.60	—	—	0.091	1.10	2.36	0.826
5.7	0.585	4.95	4.48	—	17.6	—	—	0.044	0.854	1.95	0.653

TABLE 1-3. American Standard Channels*
                    Properties for Designing

[

Designation	Area A	Depth d	Flange		Web thickness t_w	$\dfrac{d}{A_f}$	Axis X-X		
			Width b_f	Average thickness t_f			I	S	r
	In.^{2}	In.	In.	In.	In.		In.^{4}	In.^{3}	In.
C 15 × 50	14.7	15.00	3.716	0.650	0.716	6.21	404	53.8	5.24
× 40	11.8	15.00	3.520	0.650	0.520	6.56	349	46.5	5.44
× 33.9	9.96	15.00	3.400	0.650	0.400	6.79	315	42.0	5.62
C 12 × 30	8.82	12.00	3.170	0.501	0.510	7.55	162	27.0	4.29
× 25	7.35	12.00	3.047	0.501	0.387	7.85	144	24.1	4.43
× 20.7	6.09	12.00	2.942	0.501	0.282	8.13	129	21.5	4.61
C 10 × 30	8.82	10.00	3.033	0.436	0.673	7.55	103	20.7	3.42
× 25	7.35	10.00	2.886	0.436	0.526	7.94	91.2	18.2	3.52
× 20	5.88	10.00	2.739	0.436	0.379	8.36	78.9	15.8	3.66
× 15.3	4.49	10.00	2.600	0.436	0.240	8.81	67.4	13.5	3.87
C 9 × 20	5.88	9.00	2.648	0.413	0.448	8.22	60.9	13.5	3.22
× 15	4.41	9.00	2.485	0.413	0.285	8.76	51.0	11.3	3.40
× 13.4	3.94	9.00	2.433	0.413	0.233	8.95	47.9	10.6	3.48
C 8 × 18.75	5.51	8.00	2.527	0.390	0.487	8.12	44.0	11.0	2.82
× 13.75	4.04	8.00	2.343	0.390	0.303	8.75	36.1	9.03	2.99
× 11.5	3.38	8.00	2.260	0.390	0.220	9.08	32.6	8.14	3.11
C 7 × 14.75	4.33	7.00	2.299	0.366	0.419	8.31	27.2	7.78	2.51
× 12.25	3.60	7.00	2.194	0.366	0.314	8.71	24.2	6.93	2.60
× 9.8	2.87	7.00	2.090	0.366	0.210	9.14	21.3	6.08	2.72
C 6 × 13	3.83	6.00	2.157	0.343	0.437	8.10	17.4	5.80	2.13
× 10.5	3.09	6.00	2.034	0.343	0.314	8.59	15.2	5.06	2.22
× 8.2	2.40	6.00	1.920	0.343	0.200	9.10	13.1	4.38	2.34
C 5 × 9	2.64	5.00	1.885	0.320	0.325	8.29	8.90	3.56	1.83
× 6.7	1.97	5.00	1.750	0.320	0.190	8.93	7.49	3.00	1.95
C 4 × 7.25	2.13	4.00	1.721	0.296	0.321	7.84	4.59	2.29	1.47
× 5.4	1.59	4.00	1.584	0.296	0.184	8.52	3.85	1.93	1.56
C 3 × 6	1.76	3.00	1.596	0.273	0.356	6.87	2.07	1.38	1.08
× 5	1.47	3.00	1.498	0.273	0.258	7.32	1.85	1.24	1.12
× 4.1	1.21	3.00	1.410	0.273	0.170	7.78	1.66	1.10	1.17

* Taken from the 7th Edition of the *Manual of Steel Construction*. Courtesy American Institute of Steel Construction.

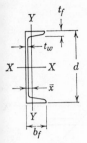

Properties for Designing

| Nominal weight per ft. | Axis Y-Y | | | $\bar{x}$ | Shear center location E_0 | Torsional constant J | Warping constant C_w |
	I	S	r				
	In.^{4}	In.^{3}	In.	In.	In.	In.^{4}	In.^{6}
50	11.0	3.78	0.867	0.799	0.941	2.66	492
40	9.23	3.36	0.886	0.778	1.03	1.46	410
33.9	8.13	3.11	0.904	0.787	1.10	1.01	358
30	5.14	2.06	0.763	0.674	0.873	0.865	151
25	4.47	1.88	0.780	0.674	0.940	0.541	131
20.7	3.88	1.73	0.799	0.698	1.01	0.371	112
30	3.94	1.65	0.669	0.649	0.705	1.22	79.5
25	3.36	1.48	0.676	0.617	0.757	0.690	68.4
20	2.8!	1.32	0.691	0.606	0.826	0.370	57.0
15.3	2.28	1.16	0.713	0.634	0.916	0.211	45.5
20	2.42	1.17	0.642	0.583	0.739	0.429	39.5
15	1.93	1.01	0.661	0.586	0.824	0.209	31.0
13.4	1.76	0.962	0.668	0.601	0.859	0.169	28.2
18.75	1.98	1.01	0.599	0.565	0.674	0.436	25.1
13.75	1.53	0.853	0.615	0.553	0.756	0.187	19.3
11.5	1.32	0.781	0.625	0.571	0.807	0.131	16.5
14.75	1.38	0.779	0.564	0.532	0.651	0.268	13.1
12.25	1.17	0.702	0.571	0.525	0.695	0.161	11.2
9.8	0.968	0.625	0.581	0.541	0.752	0.100	9.16
13	1.05	0.642	0.525	0.514	0.599	0.241	7.21
10.5	0.865	0.564	0.529	0.500	0.643	0.131	5.94
8.2	0.692	0.492	0.537	0.512	0.699	0.075	4.73
9	0.632	0.449	0.489	0.478	0.590	0.109	2.93
6.7	0.478	0.378	0.493	0.484	0.647	0.055	2.22
7.25	0.432	0.343	0.450	0.459	0.546	0.082	1.24
5.4	0.319	0.283	0.449	0.458	0.594	0.040	0.923
6	0.305	0.268	0.416	0.455	0.500	0.073	0.463
5	0.247	0.233	0.410	0.438	0.521	0.043	0.380
4.1	0.197	0.202	0.404	0.437	0.546	0.027	0.307

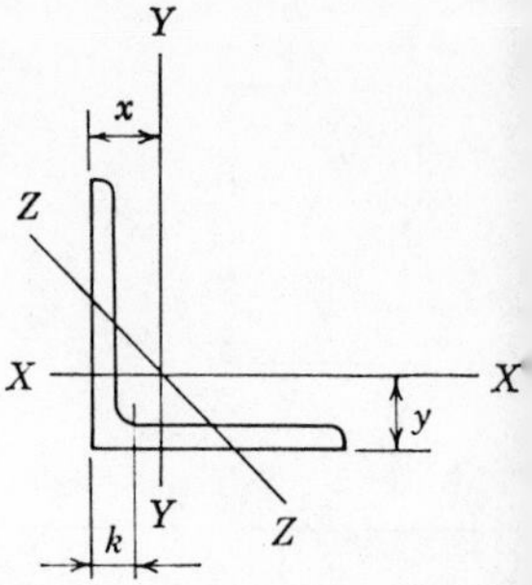

TABLE 1-4. Selected Angles, Equal Legs—Properties for Designing*

Size and thickness	k	Weight per foot	Area	Axis X-X and axis Y-Y: I	S	r	x or y	Axis Z-Z: r
In.	In.	Lb.	In.²	In.⁴	In.³	In.	In.	In.
L 8 × 8 × 1⅛	1¾	56.9	16.7	98.0	17.5	2.42	2.41	1.56
1	1⅝	51.0	15.0	89.0	15.8	2.44	2.37	1.56
¾	1⅜	38.9	11.4	69.7	12.2	2.47	2.28	1.58
L 6 × 6 × 1	1½	37.4	11.0	35.5	8.57	1.80	1.86	1.17
⅞	1⅜	33.1	9.73	31.9	7.63	1.81	1.82	1.17
¾	1¼	28.7	8.44	28.2	6.66	1.83	1.78	1.17
⅝	1⅛	24.2	7.11	24.2	5.66	1.84	1.73	1.18
L 5 × 5 × ⅞	1⅜	27.2	7.98	17.8	5.17	1.47	1.57	0.973
¾	1¼	23.6	6.94	15.7	4.53	1.51	1.52	0.975
⅝	1⅛	20.0	5.86	13.6	3.86	1.52	1.48	0.978
½	1	16.2	4.75	11.3	3.16	1.54	1.43	0.983
L 4 × 4 × ¾	1⅛	18.5	5.44	7.67	2.81	1.19	1.27	0.778
⅝	1	15.7	4.61	6.66	2.40	1.20	1.23	0.779
½	⅞	12.8	3.75	5.56	1.97	1.22	1.18	0.782
⅜	¾	9.8	2.86	4.36	1.52	1.23	1.14	0.788
5/16	11/16	8.2	2.40	3.71	1.29	1.24	1.12	0.791
L 3½ × 3½ × ½	⅞	11.1	3.25	3.64	1.49	1.06	1.06	0.683
⅜	¾	8.5	2.48	2.87	1.15	1.07	1.01	0.687
5/16	11/16	7.2	2.09	2.45	0.976	1.08	0.990	0.690
L 3 × 3 × ½	13/16	9.4	2.75	2.22	1.07	0.898	0.932	0.584
⅜	11/16	7.2	2.11	1.76	0.833	0.913	0.888	0.587
5/16	⅝	6.1	1.78	1.51	0.707	0.922	0.869	0.589
L 2½ × 2½ × ½	13/16	7.7	2.25	1.23	0.724	0.739	0.806	0.487
⅜	11/16	5.9	1.73	0.984	0.566	0.753	0.762	0.487
5/16	⅝	5.0	1.46	0.849	0.482	0.761	0.740	0.489
¼	9/16	4.1	1.19	0.703	0.394	0.769	0.717	0.491
L 2 × 2 × ⅜	⅞	4.7	1.36	0.479	0.351	0.594	0.636	0.389
5/16	13/16	3.92	1.15	0.416	0.300	0.601	0.614	0.390
¼	¾	3.19	0.938	0.348	0.247	0.609	0.592	0.391

* Compiled from data in the 7th Edition of the *Manual of Steel Construction.* Courtesy American Institute of Steel Construction.

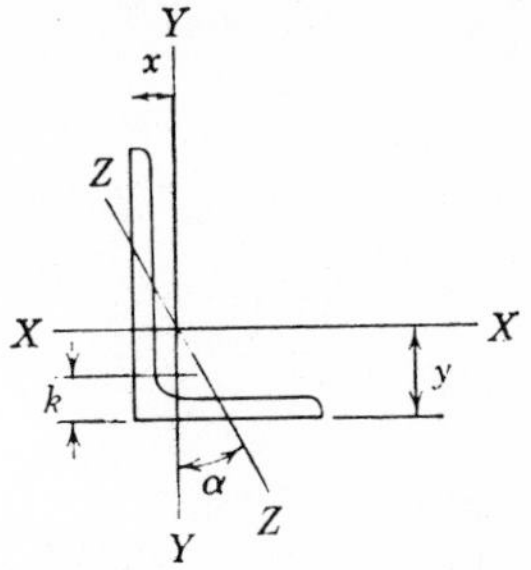

TABLE 1-5. Selected Angles, Unequal Legs—Properties for Designing*

Size and thickness	k	Weight per foot	Area	Axis X-X				Axis Y-Y				Axis Z-Z	
				I	*S*	*r*	*y*	*I*	*S*	*r*	*x*	*r*	Tan α
In.	In.	Lb.	$In.^2$	$In.^4$	$In.^3$	In.	In.	$In.^4$	$In.^3$	In.	In.	In.	
L 6×4×7/8	1 3/8	27.2	7.98	27.7	7.15	1.86	2.12	9.75	3.39	1.11	1.12	0.857	0.421
3/4	1 1/4	23.6	6.94	24.5	6.25	1.88	2.08	8.68	2.97	1.12	1.08	0.860	0.428
1/2	1	16.2	4.75	17.4	4.33	1.91	1.99	6.27	2.08	1.15	0.987	0.870	0.440
3/8	7/8	12.3	3.61	13.5	3.32	1.93	1.94	4.90	1.60	1.17	0.941	0.877	0.446
L 5×3 1/2×3/4	1 1/4	19.8	5.81	13.9	4.28	1.55	1.75	5.55	2.22	0.977	0.996	0.748	0.464
5/8	1 1/8	16.8	4.92	12.0	3.65	1.56	1.70	4.83	1.90	0.991	0.951	0.751	0.472
1/2	1	13.6	4.00	9.99	2.99	1.58	1.66	4.05	1.56	1.01	0.906	0.755	0.479
L 5×3×1/2	1	12.8	3.75	9.45	2.91	1.59	1.75	2.58	1.15	0.829	0.750	0.648	0.357
3/8	7/8	9.8	2.86	7.37	2.24	1.61	1.70	2.04	0.888	0.845	0.704	0.654	0.364
5/16	13/16	8.2	2.40	6.26	1.89	1.61	1.68	1.75	0.753	0.853	0.681	0.658	0.368
L 4×3×5/8	1 1/16	13.6	3.98	6.03	2.30	1.23	1.37	2.87	1.35	0.849	0.871	0.637	0.534
1/2	15/16	11.1	3.25	5.05	1.89	1.25	1.33	2.42	1.12	0.864	0.827	0.639	0.543
5/16	3/4	7.2	2.09	3.38	1.23	1.27	1.26	1.65	0.734	0.887	0.759	0.647	0.554
L 3 1/2×3×1/2	15/16	10.2	3.00	3.45	1.45	1.07	1.13	2.33	1.10	0.881	0.875	0.621	0.714
3/8	13/16	7.9	2.30	2.72	1.13	1.09	1.08	1.85	0.851	0.897	0.830	0.625	0.721
5/16	3/4	6.6	1.93	2.33	0.954	1.10	1.06	1.58	0.722	0.905	0.808	0.627	0.724
L 3 1/2×2 1/2×1/2	15/16	9.4	2.75	3.24	1.41	1.09	1.20	1.36	0.760	0.704	0.705	0.534	0.486
3/8	13/16	7.2	2.11	2.56	1.09	1.10	1.16	1.09	0.592	0.719	0.660	0.537	0.496
5/16	3/4	6.1	1.78	2.19	0.927	1.11	1.14	0.939	0.504	0.727	0.637	0.540	0.501
L 3×2 1/2×1/2	7/8	8.5	2.50	2.08	1.04	0.913	1.00	1.30	0.744	0.722	0.750	0.520	0.667
5/16	11/16	5.6	1.62	1.42	0.688	0.937	0.933	0.898	0.494	0.744	0.683	0.525	0.680
1/4	5/8	4.5	1.31	1.17	0.561	0.945	0.911	0.743	0.404	0.753	0.661	0.528	0.684
L 3×2×1/2	13/16	7.7	2.25	1.92	1.00	0.924	1.08	0.672	0.474	0.546	0.583	0.428	0.414
5/16	5/8	5.0	1.46	1.32	0.664	0.948	1.02	0.470	0.317	0.567	0.516	0.432	0.435
L 2 1/2×2×3/8	11/16	5.3	1.55	0.912	0.547	0.768	0.831	0.514	0.363	0.577	0.581	0.420	0.614
5/16	5/8	4.5	1.31	0.788	0.466	0.776	0.809	0.446	0.310	0.584	0.559	0.422	0.620
1/4	9/16	3.62	1.06	0.654	0.381	0.784	0.787	0.372	0.254	0.592	0.537	0.424	0.626

* Compiled from data in the 7th Edition of the *Manual of Steel Construction.* Courtesy, American Institute of Steel Construction.

2

Unit
Stresses

||||||||||||||||||||||||||||||||||||||||||||||||

2-1 General

As a prelude to discussing the strength and behavior of structural
steel under load, it will be well to establish a clear understanding of
the concept of *unit stress*. Throughout this book there will be found
many technical terms with which the reader may or may not be
familiar, depending upon the extent of his previous experience
in structural work. In any event, it is very important that these terms
and the concepts to which they apply be understood precisely.

2-2 Direct Stress

The hanger bar in Fig. 2-1a and the short block in Fig. 2-1b are both
under direct stress. The hanger, which is fastened to a ceiling plate,
supports a suspended load P acting vertically along the axis of the
bar. The load tends to elongate the bar and is called a tensile force.
The bar resists the tendency to elongate by developing an internal
tensile stress. The total internal stress is equal to the external force P
(neglecting the weight of the bar), and the *unit* tensile stress is equal

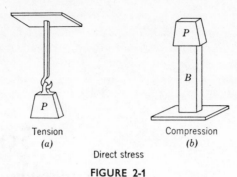

Tension
(a)

Compression
(b)

Direct stress

FIGURE 2-1

to P divided by the cross-sectional area of the hanger. If the force P equals 24,000 lb and the bar has an area of $1\frac{1}{2}$ sq in., the unit stress in the hanger is $24,000 \div 1.5 = 16,000$ psi. By using the terms

P = the external force,
A = the area of cross section,
and f_a = the axial unit stress,

this fundamental principle may be expressed in the following formula:

$$f_a = \frac{P}{A} \quad \text{or} \quad P = f_a A \quad \text{or} \quad A = \frac{P}{f_a}$$

This basic formula applies to direct stresses. It assumes that the load is axial and that the stresses are uniformly distributed over the cross section. The first form shown gives the computed unit stress when P and A are known; the second gives the allowable load when A and the allowable unit stress are known; the third gives the cross-sectional area required when the load to be carried and the allowable unit stress are known. Since the general nomenclature of the AISC Specification makes a distinction between computed and allowable unit stresses, the three forms of the direct stress formula when applied to tension situations may be written:

$$f_t = \frac{P}{A} \quad \text{or} \quad P = F_t A \quad \text{or} \quad A = \frac{P}{F_t}$$

Referring to Fig. 2-1*b*, the load *P* on the short square block exerts an axial force that tends to shorten its length. This is called a compressive force and is resisted by an internal compressive stress equal to *P*. Again, the unit compressive stress is expressed by the direct stress formula $f_a = P/A$. However, this relationship holds for short compression members only.[1] We may, therefore define a *unit stress* as an internal resistance per unit area that results from an external force. To be exact, we should, of course, consider the weight of the member. In the case of the hanger, the unit tensile stress immediately above the hook would be 16,000 psi, but the unit stress at a cross section near the ceiling plate would be slightly greater because the weight of the bar must also be supported. Similarly, the unit compressive stress at the base of the block (Fig. 2-1*b*) would be slightly greater than at a cross section just under the load.

2-3 Kinds of Stress

The three basic kinds of stress with which we are concerned are *tension*, *compression*, and *shear*. Whereas the direct tensile and compressive stresses discussed in Art. 2-2 act at right angles to the cross sections of the members considered, shearing stress acts parallel to the cross section. This is illustrated in Fig. 2-2, where two

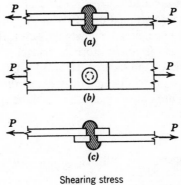

Shearing stress

FIGURE 2-2

[1] As the ratio of length to least width of compression members increases, other factors enter the problem; these are considered under column action in Chapter 10.

steel bars are shown connected by a rivet. Under the action of the forces P, it is seen that there is a tendency for the rivet to fail by shearing at the plane of contact between the two bars (Fig. 2-2c). If P is 8500 lb and the rivet has a diameter of $\frac{7}{8}$ in. with cross-sectional area of 0.6013 sq in., the unit shearing stress in the rivet is found as follows:

$$f_v = \frac{P}{A} = \frac{8500}{0.6013} = 14,130 \text{ psi}$$

It will be noted that this formula is similar to the one used for direct tension and compression (Art. 2-2). There is also an analogous assumption that the shearing stress is uniformly distributed over the cross section. However, it must be understood clearly that *the physical situations represented by the two cases are quite different.*

Another situation where shearing stress can be important is illustrated in Fig. 2-3a, which shows a beam supporting a uniformly distributed load over its length. It is evident from the sketch that the loaded beam might fail by simply dropping between the walls as indicated in Fig. 2-3b. Although the nature of beam action is such that failure would probably occur in some other manner, the tendency to drop between the walls would nevertheless be present. This tendency for one part of a beam to move vertically with respect to an adjacent part is called the *vertical shear*. It is designated by the symbol V, and the determination of its magnitude is explained in Chapter 4. It should be pointed out here, however, that the resistance to shear in this case is not uniformly distributed over the cross section of the beam; and consequently, the unit shearing stress cannot be found simply by dividing the value of the vertical shear by the cross-sectional area of the beam. This will be considered further in Art. 2-5.

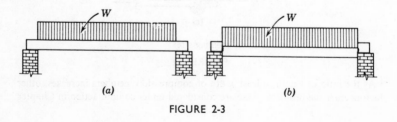

FIGURE 2-3

Problem 2-3-A. A steel bar $1\frac{1}{4}$ in. square supports a tensile load of 19,000 lb. Compute the unit tensile stress in the bar.

Problem 2-3-B. A steel rod $1\frac{1}{2}$ in. in diameter supports a tensile load of 21,000 lb. Compute the unit tensile stress.

Problem 2-3-C. A steel bar $\frac{3}{4}$ in. thick supports a tensile load of 32,000 lb. What should be its width if the allowable unit tensile stress is 22,000 psi?

Problem 2-3-D. * An L $4 \times 3 \times \frac{5}{16}$ is used as a hanger to support a load of 50,000 lb. Compute the unit tensile stress. (See Table 1-5.)

Problem 2-3-E. What should be the diameter of a steel bolt to resist a shearing stress of 13,000 lb if the allowable unit shearing stress is 22,000 psi?

2-4 Bending Stresses

The tensile and compressive stresses that accompany beam action are not direct stresses and they may not be computed by the formula $f_a = P/A$.

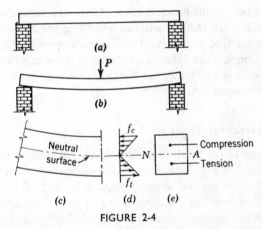

FIGURE 2-4

Figure 2-4a illustrates a rectangular beam with supports at each end. An exaggerated drawing of the shape the beam tends to assume when loaded is shown in Fig. 2-4b. The fibers[2] in the upper portion of the beam are in compression and those in the lower portion are in

---

[2] Although steel is not a fibrous material in the sense that wood is, the concept of infinitely small fibers is useful in the study of stress relationships within any material.

tension. Figure 2-4c is an enlarged detail representing a segment of
the beam cut at a section near the middle of the span. The line
marked *neutral surface* locates the plane between the upper and
lower surfaces above which the stresses are compressive and below
which they are tensile. The line in which the neutral surface cuts the
cross section of the beam is called the *neutral axis* (Fig. 2-4e). In
beams of rectangular or other symmetrical cross section (such as
I-beams and wide flange shapes), the neutral surface occurs midway
between the top and bottom surfaces.

Figure 2-4d shows the distribution of the stresses at the cut
section. The maximum tensile and compressive stresses that occur at
the upper and lower surfaces of the beam are called *extreme fiber
stresses* or simply *bending stresses* (f_b). The stresses decrease in
magnitude toward the neutral surface, where they become zero. If
the magnitude of the bending stress at the extreme fibers does not
exceed certain limits, the stresses over the cross section are directly
proportional to their distances from the neutral surface. In the
design of beams, the problem is to select a structural shape that will
sustain the given loading without exceeding the allowable bending
stress for the type of structural steel being employed. The expression
used to compute the value of the bending stress in either tension or
compression is known as the *beam formula* or the *flexure formula*
and is developed in Chapter 5.

2-5 Horizontal Shear

In addition to the bending stresses developed when a beam deflects
as indicated in Fig. 2-4b, there is also a horizontal shearing stress
related to the vertical shear discussed in Art. 2-3. Figure 2-5a
represents a beam of the same size and span length as the one in
Fig. 2-4b but instead made up of three independent strips. If subjected to the same loading, its deflection would be greater than that of
the solid beam, and slipping would occur along the surfaces of contact
between the independent strips (Fig. 2-5a). This same tendency is
present in the solid beam, but the action is restrained by its resistance
to *horizontal shear*.

It can be shown that at any point in a beam the intensity of the
horizontal shear is equal to the intensity of the vertical shear. The

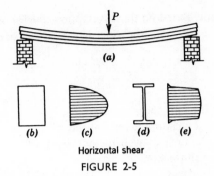

Horizontal shear
FIGURE 2-5

shearing stresses, however, are not distributed uniformly over the cross-sectional area.[3] For rectangular beam cross sections, the stresses vary as shown by the parabola (Fig. 2-5c), the length of a horizontal line representing the magnitude of the stress at that point in the depth of the beam. From the figure, it is apparent that the maximum unit shearing stress occurs at the neutral surface. Its magnitude is 1.5 times the average unit stress and may be computed by the equation

$$v = \frac{3}{2} \times \frac{V}{A}$$

in which v = the maximum unit shearing stress, either horizontal or vertical,

V = the maximum vertical shear,

and A = the area of the rectangular section.

This equation applies only to rectangular beams and, consequently, must be modified when investigating wide flange and I-beam sections where most of the material lies in the flanges. Figure 2-5e indicates the distribution of the shearing stresses in these structural shapes. Since the maximum shearing stress occurs at the neutral surface and the value at the extreme fibers is zero, the flanges have little influence on resistance to shear. It is, therefore, customary to ignore the material in the flanges and to consider only the web as

[3] See Horizontal Shear in *Simplified Mechanics and Strength of Materials*, Second Edition, by Harry Parker, John Wiley & Sons, New York.

resisting the shear. Based on this assumption, the following approximate expression for unit shearing stress is the one customarily employed:

$$f_v = \frac{V}{A_w}$$

in which f_v = the unit shearing stress,

V = the maximum vertical shear,

and A_w = the area of web (actual depth of section times the web thickness).

Here, of course, the value of f_v is really the *average* unit shearing stress over the area of the web. The fact that the maximum unit stress is somewhat greater than the average value (Fig. 2-5e) is handled in practice by assigning a value for the *allowable* shearing stress low enough to compensate for this difference.

Example.   A W 12 × 36 is used as a beam subjected to a vertical shear at the supports (Fig. 2-3) of 50,000 lb. Find the value of the unit shearing stress.

Solution: Referring to Table 1-1, we find that the depth of this section is 12.24 in. and the web thickness is 0.305 in. Then the area of the web is 12.24 × 0.305 or 3.73 sq in. Therefore,

$$f_v = \frac{V}{A_w} = \frac{50,000}{3.73} = 13,400 \text{ psi}$$

Problem 2-5-A. If the maximum vertical shear on an S 18 × 70 is 100,000 lb, determine the value of the unit shearing stress.

Problem 2-5-B. If the maximum vertical shear on a W 18 × 70 is 100,000 lb, determine the value of the unit shearing stress.

2-6 Elastic Limit, Yield Point, and Ultimate Strength

When a member is subjected to a load, there is always an accompanying change in length or shape of the member. Such a change is called a *deformation* or *strain*. For compression and tension, the deformations are, respectively, a shortening and lengthening. The deformation accompanying bending is called *deflection*.

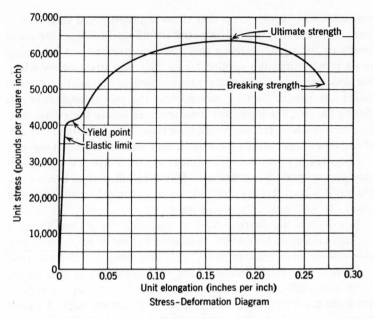

Stress–Deformation Diagram

FIGURE 2-6

In order to investigate the relation between the magnitudes of deformation and the accompanying stresses, it is convenient to record them on a sheet of graph paper. Figure 2-6 illustrates such a chart. As an example, a structural steel bar was placed in a testing machine and subjected to a tensile test. To record the stresses and deformations, a vertical scale, representing the unit stresses in pounds per square inch, was laid off, and a horizontal scale, shown at the bottom of the chart, was made for the deformations. The latter scale indicates the deformations of the specimen in inches per inch of length. At various stresses, points were plotted on the chart to indicate the accompanying deformations. When the bar was finally stressed to rupture, the different points were connected; the curve, represented by the heavy black line, presents graphically the results of the test.

Note that up to a unit stress of about 38,000 psi the curve is a straight line. This is important, for it shows that the deformations up to this stress are directly proportional to the applied loads, a verification of Hooke's law that stress is proportional to deformation. It is seen that beyond this stress of about 38,000 psi the curve bends to the right, showing that the deformations are increasing more rapidly than the applied loads. This unit stress below which the deformations are directly proportional to the unit stresses is called the *elastic limit* or, sometimes, the *proportional limit*. If, during the test, the loads had been removed before the elastic limit had been reached, the bar would have returned to its original length. If the unit stress exceeds the elastic limit and the load is removed, the bar will not return to its original length. There is a permanent lengthening, and this increase in length is called a *permanent set*. Hence, it is seen that the unit stresses in structural members should always be well within the elastic limit.

Just beyond the elastic limit it is seen that for a short distance the curve is almost horizontal, indicating a *slight amount of deformation without an increase in stress*. This unit stress is called the *yield point*. In testing steel specimens, the yield point may be determined more accurately than the elastic limit; the two stresses, however, are quite close together. With respect to structural steel, the yield point is a particularly important unit stress. The AISC Specification, in giving the various allowable unit stresses, designates these stresses as fractions of the yield point.

Beyond the yield point, the curve begins to flatten out until the greatest stress is reached. This is the *ultimate strength*. It occurs at or immediately before rupture. It should be noted that the deformations increase rapidly beyond the yield point. The specimen not only elongates, but there is also a reduction in the cross-sectional area. Failure begins at the ultimate strength, the greatest unit stress reached, and rupture occurs at the point marked *breaking strength*.

2-7 Modulus of Elasticity

The *modulus of elasticity* of a material is a number that indicates its degree of *stiffness*. A material is said to be stiff if its deformation is relatively small when the unit stresses are high. As an example, a steel rod 1 sq in. in cross-sectional area and 10 ft in length will elongate about 0.008 in. under a tensile load of 2000 lb. But a piece of wood of the same dimensions and with the same tensile load will stretch about 0.24 in., or nearly ¼ in. We say that the steel is stiffer than the wood because, for the same unit stress, the deformation is not so great.

It is quite a simple matter to compute the deformations for tensile and compressive loads. It is important to remember, however, that for such computations we must be careful to see that the unit stress does not exceed the elastic limit of the material. In Fig. 2-6, it was noted that the deformation curve up to the elastic limit was a straight line, since the deformations and stresses were directly proportional. This proportion may be expressed as *the unit stress divided by the unit deformation*. The name for this ratio is the *modulus of elasticity*. It is represented by the letter *E*, expressed in units of pounds per square inch, and has the same value for both compression and tension for most structural materials.

E = the modulus of elasticity (psi or ksi)
P = the applied force in pounds or kips
l = the length of the member in inches
A = the cross-sectional area of the member in square inches
e = the total deformation of the member in inches
s = the unit deformation in inches per inch
f = the unit stress in pounds per square inch

Since, by definition, the modulus of elasticity is the unit stress divided by the unit deformation,

$$E = \frac{f}{s} \quad \text{or} \quad E = \frac{P/A}{e/l} \quad \text{or} \quad E = \frac{Pl}{Ae} \quad \text{or} \quad e = \frac{Pl}{AE}$$

There are five terms in this equation, and if any four are known the fifth term may be found. In computing the deformation by the use of this formula, we must, of course, know the modulus of elasticity of the material. For structural steel, $E = 29,000,000$ psi. Depending on the species,[4] E for wood varies from 1,000,000 to 1,760,000 psi. Bear in mind that the foregoing formula is valid only when the unit stress lies within the elastic limit of the material. Let us try an example.

---

[4] See *Simplified Design of Structural Timber*, Second Edition, by Harry Parker, John Wiley & Sons, New York.

Example. A bar 1 × 2 in. in cross section and 20 ft long is made from structural steel having a yield point of 42,000 psi. Compute its total elongation under a tensile force of 36,000 lb.

Solution: The bar has an area of 2 × 1, or 2 sq in., and since the applied load is 36,000 lb, the unit stress developed is 36,000/2 or 18,000 psi. This unit stress is so much lower than the yield point (which is only slightly higher than the elastic limit) that the above formula is applicable. Then

$$e = \frac{Pl}{AE} \quad \text{or} \quad e = \frac{36,000 \times 20 \times 12}{2 \times 29,000,000} = 0.149 \text{ in.}$$

is the total elongation. Note in the numerator of this equation (20 × 12). This is *l*, the length of the member, *in inches.*

Problem 2-7-A. In the first paragraph of Art. 2-7, it was stated that a steel rod of certain dimensions and a unit stress of 2000 psi would elongate about 0.008 in. How do we know this is true?

Problem 2-7-B.* A steel rod 1 in. in diameter and 8 in. long is placed in a testing machine and tested for tension. Under a load of 15,000 lb, the deformation is found to be 0.0052 in. Compute the modulus of elasticity.

2-8 Factor of Safety

The strength of a structural member must be greater than the force it is required to resist. The uncertainties with regard to determination of actual loading and uniformity in quality of materials require that some reserve strength be built into the design. This degree of reserve strength is the *factor of safety.* Although there is no general agreement on an exact definition of this term, the following discussion will serve to fix the concept in mind.

Consider a structural steel that has an ultimate tensile strength of 70,000 psi, a yield point stress of 36,000 psi, and an allowable unit tensile stress of 22,000 psi. If the factor of safety is defined as the ratio between the ultimate strength and the allowable stress, its value is 70,000/22,000 or 3.18. On the other hand, if it is defined as the ratio of the yield point stress to the allowable stress, its value is 36,000/22,000 or 1.64. This is a considerable variation, and since failure of a structural member begins when it is stressed beyond the elastic limit, the higher value may be misleading. It must be borne in mind that stresses beyond the elastic limit produce deformations that are *permanent* and thus change the shape of a member, even though there may not be danger of collapse.

2-9 Allowable Stresses

In our discussion of unit stresses we made a distinction between the computed value of a particular kind of stress and the allowable value. It is the latter that is controlled by the requirements of local building codes. As noted earlier, there is some variation among these codes, and the allowable unit stresses in some are more conservative than in others. The user of this book should obtain a copy of the building requirements that govern construction in his community and familiarize himself with those items that pertain to structural steel.

As an example of the differences in building codes, consider bending. Older codes that have not been kept up to date may give 22,000 psi as the allowable extreme fiber stress for certain members in flexure, whereas 24,000 psi is the value now generally accepted for the grade of structural steel commonly used in building construction. A table of allowable unit stresses (Table 3-2) is given in the following chapter. Referring to the table, it will be noted that stress values are expressed in kips per square inch (ksi) rather than in pounds per square inch (psi). This practice follows that adopted in the 1969 AISC Specification.

Review Problems

Problem 2-9-A.* If the allowable unit tensile stress of steel is 22,000 psi, will a $1\frac{1}{2}$ in. diameter rod support safely a suspended load of 50,000 lb?

Problem 2-9-B. Assume the diameter of the rivet shown in Fig. 2-2 to be $\frac{3}{4}$ in. If the allowable unit shearing stress of the rivet steel is 15,000 psi, compute the maximum value of *P*.

Problem 2-9-C. Compute the maximum vertical shear that a W 10 × 21 can resist if the allowable unit shearing stress is 14,500 psi. *Note:* For the dimensions of wide flange beams, see Table 1-1.

Problem 2-9-D. Compute the maximum vertical shear that an S 10 × 25.4 can resist if the allowing unit shearing stress is 14,500 psi.

Problem 2-9-E.* A steel rod 1 in. in diameter and 18 ft 6 in. long is subjected to a tensile load of 15,500 lb. Compute the total elongation.

Problem 2-9-F. A steel rod ½ in. in diameter and 9 in. long is placed in a testing machine and subjected to a tensile load of 4000 lb. Its total elongation under this loading is 0.007 in. Find the unit stress in the rod under these conditions.

3

Structural
Steel

‖‖

3-1 Grades of Structural Steel

Steel meeting the requirements of the American Society for Testing
and Materials Specification A36 is the grade of structural steel
commonly used for building construction. It is required to have an
ultimate tensile strength of 58 to 80 ksi and a minimum yield point of
36 ksi. It may be used for bolted, riveted, and welded fabrication.

In addition to A36 as the all-purpose carbon grade steel, other
special steels are permitted by the 1969 AISC Specification. Several
of these are listed below.

Structural Steel, ASTM A36
Structural Steel with 42,000 psi Minimum Yield Point, ASTM 529
High-Strength Structural Steel, ASTM A440
High-Strength Low-Alloy Structural Manganese Vanadium Steel,
ASTM A441
*High-Strength Low-Alloy Columbium-Vanadium Steels of Struc-
tural Quality*, ASTM A572
High-Strength Low-Alloy Structural Steel, ASTM A242
*High-Strength Low-Alloy Structural Steel with 50,000 psi Minimum
Yield Point to 4 in. Thick*, ASTM A588.

39

TABLE 3-1. Structural Steels for Buildings—1969 AISC Specification

| Steel type | ASTM number | Strength | | Fabrication* |
		Thickness (in.)	Yield F_y (ksi)	
Carbon	A36		36	B R W
	A529		42	B R W
High-Strength		$1\frac{1}{2}$ to 4	42	
	A440	$\frac{3}{4}$ to $1\frac{1}{2}$	46	B R
		to $\frac{3}{4}$	50	
High-Strength Low-Alloy	A441	4 to 8	40	B R W
		$1\frac{1}{2}$ to 4	42	
		$\frac{3}{4}$ to $1\frac{1}{2}$	46	
		to $\frac{3}{4}$	50	
	A572		42	B R W
Corrosion-Resistant High-Strength Low-Alloy	A242	$1\frac{1}{2}$ to 4	42	B R W
		$\frac{3}{4}$ to $1\frac{1}{2}$	46	
		to $\frac{3}{4}$	50	
	A588	5 to 8	42	B R W
		4 to 5	46	
		to 4	50	

* B, R, and W indicate that the steel is suitable for fabrication by means of bolting, riveting, or welding, respectively.

All of the above steels are suitable for welding except A440, which is not recommended for welding by the AISC.

Table 3-1 summarizes certain properties of these structural steels. Note in particular that F_y, the minimum yield stress, is given for each type of steel. This stress is important because allowable unit stresses are specified in terms of the minimum yield stress for a particular type of steel.

3-2 Structural Carbon Steels

ASTM A36 steel has supplanted ASTM A7, which was the basic steel for structural purposes prior to the promulgation of the 1963 AISC Specification. The A7 steel was employed primarily for riveted fabrication; with the advent of welded construction, its use decreased in favor of other steels that better satisfy requirements for weldability. Its yield point was 33 ksi. ASTM A373 was developed as a more weldable steel than A7, but both of these structural grades have been withdrawn from production. ASTM A36 fully satisfies requirements for a weldable steel, and its higher yield point (36 ksi) permits higher allowable unit stresses.

A 529 Steel has a yield point of 42 ksi. Its availability is limited to the lighter-weight rolled shapes and plates and bars up to ½ in. thickness.

3-3 High-Strength Steels

Reference to Table 3-1 shows that the following steels come under the high-strength classification: A440, A441, A572, A242, and A588. The yield points for these steels are higher than those of structural carbon steel; consequently, their allowable unit stresses are higher as well. It will also be noted from Table 3-1 that the thickness of the material is a factor in establishing the yield stress. As noted earlier, A440 steel is used for bolted and riveted construction and is not recommended for welded fabrication by the AISC.

A441 is a high-strength steel intended primarily for use in welded construction; A572 is suitable for bolting, riveting, or welding.

3-4   Corrosion-Resistant Steels

The corrosion-resistant high-strength low-alloy steels are A242 and A588. Because of their chemical composition, these steels, in addition to their greater strengths, exhibit higher degrees of atmospheric corrosion resistance. For A242, this resistance is 4 to 6 times that of structural carbon steel; for A588, the resistance is about 8 times as great as for A440 and A441. Several brands of A588 steel are now available, each of which represents the proprietary product of a different manufacturer. These are the steels now being used in the bare (uncoated) condition for exposed-frame construction. Exposure to normal atmosphere causes formation of an oxide on the surface; this oxide adheres tightly and protects the steel from further oxidation. These steels weather to a deep rust color and do not require painting.

3-5   Allowable Stresses for Structural Steel

For the structural steels discussed in the preceding articles, The AISC Specification expresses the allowable unit stresses in terms of F_y, the yield stress. Thus, the allowable unit bending stress in both tension and compression for certain groups of beam sections is established by the relationship $F_b = 0.66F_y$. Similarly, the allowable direct tensile stress is determined by $F_t = 0.60F_y$. For A36 steel, these two allowable stresses are

$$F_b = 0.66F_y = 0.66 \times 36 = 24 \text{ ksi} \quad \left(\begin{array}{l} \text{Rounded off values.} \\ \text{See Note, Table 3-2.} \end{array} \right)$$
$$F_t = 0.60F_y = 0.60 \times 36 = 22 \text{ ksi}$$

Selected allowable unit stresses used in design are presented in Table 3-2. This is not a complete list; only the stresses used in the examples and problems given in this book are shown. A complete tabulation is given in Section 1.5 and Appendix A of the 1969 AISC Specification, which is printed in full in the Seventh Edition of the AISC Manual. It will be noted that the first column of Table 3-2 contains several qualifying terms, such as net section, gross section, and compact beams, that have not been introduced in this book so far. These will be identified in subsequent chapters dealing with the design of members to which they apply.

TABLE 3-2. Allowable Unit Stresses for Structural Steel

	AISC Specification 1969	A36	A242, A440, A441		
			Section thickness		
			Over 1½ in. and up to 4 in., incl.	Over ¾ in. and up to 1½ in., incl.	Up to ¾ in., incl.
Note: Stresses are given in kips per sq in. (ksi) and are rounded off to conform with the values given in Appendix A of the 1969 AISC Specification		$F_y = 36.0$	$F_y = 42.0$	$F_y = 46.0$	$F_y = 50.0$
Tension					
Tension on net section, expect at pin holes	$F_t = 0.60F_y$	22.0	25.2	27.5	30.0
Tension on net section at pin holes	$F_t = 0.45F_y$	16.2	19.0	20.5	22.5
Shear					
Shear on gross section	$F_v = 0.40F_y$	14.5	17.0	18.5	20.5
Compression					
See Chapter 10					
Bending					
Tension and compression on extreme fibers of compact members braced laterally, symmetrical about and loaded in the plane of their minor axis	$F_b = 0.66F_y$	24.0	28.0	30.5	33.0
Tension and compression on extreme fibers of unsymmetrical rolled shapes, except channels, continuously braced in the region under compression stress	$F_b = 0.60F_y$	22.0	25.2	27.5	30.0
Tension on extreme fibers of other rolled shapes, built-up members and plate girders	$F_b = 0.60F_y$	22.0	25.2	27.5	30.0
Compression on extreme fibers of channels	See	22.0	Art. 8-8		
Tension and compression on extreme fibers of rectangular bearing plates	$F_b = 0.75F_y$	27.0	31.5	34.5	37.5
Bearing					
Bearing on contact area of milled surfaces	$F_p = 0.90F_y$	33.0	38.0	41.5	45.0

3-6   Grades of Steel for Connectors

The AISC Specification designates the type of material permitted in bolted, riveted, and welded connections. High-strength bolts must meet one of the following American Society for Testing and Materials specifications: ASTM A325, ASTM A449, ASTM A490. Bolts other than high-strength must conform to ASTM A307.

Rivets must meet the provisions of the *Specification for Structural Rivets*, ASTM A502, Grade 1 or Grade 2.

Filler metal for welding is required to meet the specifications of the American Welding Society for the type of welding process and the type of steel being employed. There are five such specifications: AWS A5.1, AWS A5.5, AWS A5.17, AWS A5.18, AWS A5.20.

Allowable stresses for bolts and rivets are discussed in Chapter 11 under design procedures for bolted and riveted connections; allowable stresses in welds are considered in Chapter 12, Welded Connections.

4

Reactions,
Moments,
and Shear

||||||||||||||||||||||||||||||||||||||||||||||||

4-1 General

A beam is a structural member that resists transverse loads. Sometimes it is defined as a structural member on which the applied loads
are perpendicular to its longitudinal axis. The loads on a beam tend
to *bend* it, and we say the member is in *flexure* or *bending*. The
supports for beams are usually at or near the ends, and the supporting
upward forces are called *reactions*.

The term *girder* is generally applied to the larger beams. All
girders are beams insofar as their structural action is concerned. In
the floor framing of a building, the larger beams that support the
smaller beams or joists are called girders.

4-2 Types of Beam

There are several types of beam, all of which are identified by the
number, kind, and position of the supports. Figure 4-1 shows

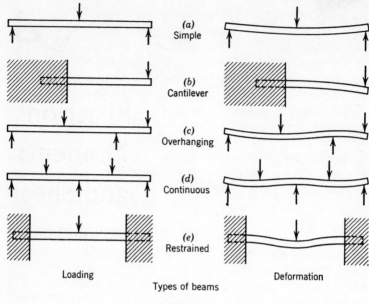

Loading   Deformation

Types of beams

FIGURE 4-1

diagrammatically the different types and also the shape each beam tends to assume as it deforms under the loading.

A *simple beam* rests on a support at each end, the beam ends being free to rotate. A large percentage of the beams in steel-frame buildings are designed as simple beams (Fig. 4-1*a*). When there is no restraint against rotation at the supports, we say that the beam is *simply supported.*

A *cantilever beam* projects beyond its support. A beam embedded in a wall and extending beyond the face of the wall is a typical example. This is illustrated in Fig. 4-1*b*. Such a beam is said to be *fixed* or *restrained* at the support.

An *overhanging beam* is a beam whose end or ends project beyond its supports. Figure 4-1*c* indicates an overhanging beam. The projecting end is a cantilever in which the stresses are similar to those in the beam shown in Fig. 4-1*b*. It will be seen later, however, that the stresses in the portion of the beam between the supports are quite different from the stresses in the simple beam shown in Fig. 4-1*a*.

A *continuous beam* is a beam supported on more than two supports (Fig. 4-1*d*). Continuous beams are frequently used in welded steel floor framing. A *restrained beam* has one or both ends restrained or *fixed* against rotation (Fig. 4-1*e*). Fixed-end conditions occur when the end of a beam is rigidly connected to its supporting member.

When designing beams, it is advantageous to make a diagram showing the loading conditions and to sketch directly below this the shape the beam tends to assume when it bends. The ability to visualize this deformation curve is very important since, as we shall see later, the curve indicates changes in the character of the bending stresses along the beam's length.

4-3 Loading

The loads supported by beams are classified as either *concentrated* or *distributed*. A concentrated load is one that extends over so small a portion of the beam length that it may be assumed to act at a point, as indicated in the diagrams of Fig. 4-1. A girder in a building receives concentrated loads at the points where the floor beams frame into it (Fig. 4-2). Also, the load exerted by a column that rests on a beam or girder is a concentrated load. Actually this load extends over a short length of the beam, represented by the width of the column base. However, for practical purposes, the load is considered to act at the axis of the column. Similarly, a beam resting on a masonry wall produces a downward force that is resisted by the upward reaction. The reaction is distributed over the length of the beam (or its bearing plate) that rests on the wall, but the reaction may be considered to act at the midpoint of the beam's bearing area on the wall.

A distributed load is one that extends over a substantial portion or the entire length of a beam. Most distributed loads are *uniformly distributed*; that is, they have a uniform magnitude for each unit of length, such as pounds per linear foot or kips per linear foot.

Figure 4-2*a* represents a portion of a floor framing plan; the diagonal crosshatching shows the area supported by one of the beams. This area is 8 ft (the sum of half the distances to the next beam on each side) multiplied by the span length of 20 ft. The beam

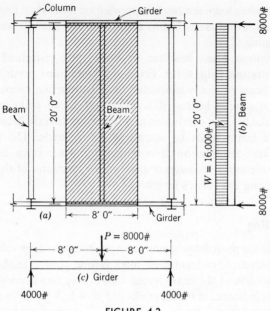

FIGURE 4-2

is supported at each end by girders spanning between the columns. Imagine the floor load to be 100 lb per sq ft, including the weight of the construction. The total load on the beam is then 8 × 20 × 100 = 16,000 lb or 16 kips. This total uniformly distributed load is designated by W, and we say $W = 16$ kips. For the same loading, we might have said that the uniformly distributed load was 800 lb per lin ft or 0.8 kips per lin ft. The distinction between W and w is that W represents the *total uniformly distributed load* and w represents the *uniformly distributed load per linear foot*.

Since the beam is symmetrically loaded, both reactions are the same and are equal to 16,000/2 or 8000 lb. Then each girder receives a concentrated load of 8000 lb at the center of its span. The conventional representations of the beam and girder load diagrams are shown in Figs. 4-2b and 4-2c respectively.

In practice, most of the distributed loads on beams are uniformly distributed and extend over the entire length of the beam. If a beam has the same cross section throughout its length, as is usually the case, its own weight constitutes a uniformly distributed load; this load must be considered in the design of the beam. Sometimes the uniform load may extend over only a portion of the beam length, as indicated in Fig. 4-15a.

4-4 Moments

The *moment* of a force is its tendency to cause rotation about a given point or axis. The magnitude of a moment is the magnitude of the force multiplied by the perpendicular distance from its line of action to the point about which the moment is taken. Moments are expressed in compound units such as foot-pounds and inch-pounds, or kip-feet and kip-inches. It is extremely important to remember that we cannot compute the moment of a force without having in mind *the point or axis about which the force tends to cause rotation.*

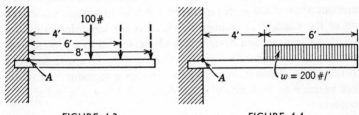

FIGURE 4-3   FIGURE 4-4

Figure 4-3 represents a cantilever beam with a concentrated load of 100 lb placed 4 ft from the fact of the wall. In this position, the moment of the force *about point A* is 4 × 100 or 400 ft-lb. If the load is moved 2 ft farther to the right, the moment of the force *about point A* is 600 ft-lb. When the load is moved to the end of the beam, the moment about the same point A is 800 ft-lb.[1]

Figure 4-4 shows a cantilever beam 10 ft 0 in. in length with a uniformly distributed load of 200 lb *per lin ft* extending over a

---

[1] To resolve any confusion about the compound units in which moments are expressed, the reader should satisfy himself that if the "moment arms" in Fig. 4-3 were stated in inches instead of feet, the three computed moments would be 4800 in-lb, 7200 in-lb, and 9600 in-lb respectively. They could also be written 4.8 kip-in, 7.2 kip-in, and 9.6 kip-in.

length of 6 ft at the position indicated. Since $w = 200$ lb per ft, W (the total distributed load) $= 200 \times 6$ or 1200 lb. In computing the moment of a distributed load about some specific point, we consider the load to act as its center of gravity. For the distributed load in Fig. 4-4, the center of gravity is 3 ft from the right end of the beam and 7 ft from the face of the wall. For instance, suppose we are asked to find the moment of this distributed load *about point A*, a point at the face of the wall. Then, as the center of gravity of the load is 7 ft from point A, the moment of the force is 1200×7 or 8400 ft-lb. The distance 7 ft is called the *lever arm* or the *moment arm*. Sometimes we refer to the point A as the *center of moments*. The important thing to remember is that in computing a moment we must have in mind the particular point about which the moment is taken.

Shortly we shall write an *equation of moments* in which will occur the moments of several forces. In writing such an equation, the moment of each force must be taken *about the same point*. Careful attention to identifying the center of moments and the lengths of the moment arms of the forces involved will lead to a ready understanding of the concept of moments, which is fundamental to much of structural analysis and design. It is also essential to a precise understanding of moments that their values always be expressed in the correct units; if one does not know whether a moment magnitude is foot-pounds or inch-pounds, he does not know the value of the moment.

4-5 Laws of Equilibrium

When a body is acted on by a number of forces, each force tends to move the body. If the forces are of such magnitude and position that their combined effect produces no motion of the body, the forces are said to be in *equilibrium*.

The three fundamental laws of equilibrium follow:

1. The algebraic sum of all the horizontal forces equals zero.
2. The algebraic sum of all the vertical forces equals zero.
3. The algebraic sum of the moments of all the forces about any point equals zero.

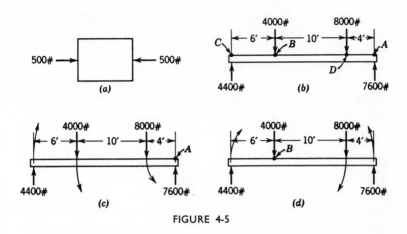

FIGURE 4-5

Figure 4-5a represents a body subjected to two equal horizontal forces having the same line of action. Obviously the forces are in equilibrium, one balancing the other. Call the force acting toward the right positive and the force acting toward the left negative. Then, in accordance with the first law of equilibrium,

$$500 - 500 = 0 \qquad \text{or} \qquad 500 \text{ lb} = 500 \text{ lb}$$

Figure 4-5b represents a simple beam. Four vertical forces act on this beam and are in equilibrium. The two downward forces, or loads, are 4000 and 8000 lb. The two upward forces, reactions, are 4400 and 7600 lb. If we call the upward forces positive and the downward forces negative, by the second law of equilibrium

$$4400 + 7600 - 4000 - 8000 = 0$$

or

$$12,000 \text{ lb} = 12,000 \text{ lb}$$

Accordingly, if the forces are in equilibrium, we may say the sum of the downward forces equals the sum of the upward forces or the sum of the loads equals the sum of the reactions. If, for instance, we had known the magnitude of the loads and also the magnitude of the left reaction, we could have written

$$4000 + 8000 = 4400 + R_2 \qquad \text{or} \qquad R_2 = 7600 \text{ lb}$$

Now consider the third law of equilibrium. Let us take the point marked A, on the line of action of the right support, and write an equation of moments. See Fig. 4-5c. We will consider the moments of the forces tending to cause clockwise rotation as positive and those tending to cause counterclockwise rotation as negative. The directions of the rotations are indicated by the curved arrow lines. Then

$$(4400 \times 20) - (4000 \times 14) - (8000 \times 4) = 0$$

or

$$(4400 \times 20) = (4000 \times 14) + (8000 \times 4)$$

or

$$88{,}000 \text{ ft-lb} = 88{,}000 \text{ ft-lb}$$

It is now apparent that in accordance with the third law of equilibrium the sum of the moments tending to cause clockwise rotation equals the sum of the moments tending to cause counterclockwise rotation.

Note carefully that in writing the equation of moments the moment of each force is taken with respect to the same point, point A in this case. Attention is also called to the moment of the force 7600 lb about point A. Since the line of action of this force passes through point A, it can cause no rotation about the point. The lever arm of the moment is zero, or $7600 \times 0 = 0$. We may say that the moment of a force about a point in its line of action is zero. When writing an equation of moments, then, it will be unnecessary to consider the moment of a force *if the center of moments is on the line of action of the force.*

According to the third law of equilibrium, we may consider *any* point. Let us take point B, Fig. 4-5d, and see if the law holds. Then

$$(4400 \times 6) + (8000 \times 10) = (7600 \times 14)$$
$$26{,}400 + 80{,}000 = 106{,}400$$
$$106{,}400 \text{ ft-lb} = 106{,}400 \text{ ft-lb}$$

Here again we have omitted writing the moment of the force 4000 about point B, for we know that $4000 \times 0 = 0$.

Another example that will serve to fix in mind the principle of moments is indicated in Fig. 4-6. The beam is 9 ft long with a single pedestal-type support located 6 ft from the left end. What should be

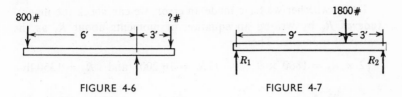

FIGURE 4-6   FIGURE 4-7

the magnitude of the load at the right end of the beam to produce equilibrium? Let us call this unknown force x and consider the point of support as the center of moments. Then, since the moment of the force tending to cause clockwise rotation must equal the moment of the force tending to produce counterclockwise rotation, we may write

$$3 \times x = 800 \times 6 \quad \text{or} \quad 3x = 4800 \quad \text{or} \quad x = 1600 \text{ lb.}$$

The reaction supplied by the single support is $800 + 1600$ or 2400 lb.

Problem 4-5-A. Write the two equations of moments for the four forces shown in Fig. 4-5b, taking points C and D as the centers of moments.

4-6 Determination of Reactions

As noted earlier, *reactions* are the forces at the supports of beams that hold the loads in equilibrium. We have seen that the sum of the loads is equal to the sum of the reactions. It is the third law of equilibrium that enables us to compute the magnitudes of the reactions. Throughout this book, the left reaction is denoted R_1 and the right reaction is denoted R_2. Another conventional designation frequently used is R_L and R_R.

A simple beam 12 ft in length has a concentrated load of 1800 lb 9 ft from the left support, as shown in Fig. 4-7. Let us compute the reactions. To do this, we simply write the sum of moments tending to cause clockwise rotation and equate it to the sum of the moments tending to cause counterclockwise rotation. Then, taking R_2 (the right reaction) as the center of moments,

$$12 \times R_1 = 1800 \times 3 \quad \text{or} \quad 12R_1 = 5400 \quad \text{and} \quad R_1 = 450 \text{ lb}$$

To find R_2, we know that $R_1 + R_2 = 1800$ lb or $450 + R_2 = 1800$. Therefore,

$$R_2 = 1800 - 450 \quad \text{and} \quad R_2 = 1350 \text{ lb}$$

To see whether we have made an error, we can check the magnitude of R_2 by writing an equation of moments about R_1 as the center of moments. Then

$$12 \times R_2 = 1800 \times 9 \quad \text{or} \quad 12R_2 = 16{,}200 \quad \text{and} \quad R_2 = 1350 \text{ lb}$$

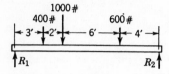   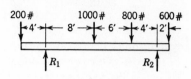

FIGURE 4-8   FIGURE 4-9

Example 1. The simple beam in Fig. 4-8 is 15 ft long and has three concentrated loads as indicated. Compute the reactions.

Solution: Taking R_2 as the center of moments,

$$15R_1 = (400 \times 12) + (1000 \times 10) + (600 \times 4)$$
$$15R_1 = 17{,}200 \quad \text{and} \quad R_1 = 1146.7 \text{ lb}$$

To compute R_2, take R_1 as the center of moments. Then

$$15R_2 = (400 \times 3) + (1000 \times 5) + (600 \times 11)$$
$$15R_2 = 12{,}800 \quad \text{and} \quad R_2 = 853.3 \text{ lb}$$

To check the results,[2]

$$400 + 1000 + 600 = 1146.7 + 853.3$$
$$2000 \text{ lb} = 2000 \text{ lb}$$

Example 2. Compute the reactions for the overhanging beam shown in Fig. 4-9.

Solution: Select R_2 as the center of moments and equate the sum of the moments tending to produce clockwise rotation to the sum of the moments tending to produce counterclockwise rotation. Then

$$18R_1 + (600 \times 2) = (200 \times 22) + (1000 \times 10) + (800 \times 4)$$
$$18R_1 = 16{,}400 \quad \text{and} \quad R_1 = 911.1 \text{ lb}$$

---

[2] It is not usual to carry the values of reactions to the nearest pound or tenth of a pound. Three or at most four significant figures (slide-rule accuracy) is sufficient in structural work. On this basis, R_1 and R_2 of the example would have values of 1147 lb and 853 lb respectively. Nevertheless, to clarify the explanation in some of the examples, the numerical work has been extended.

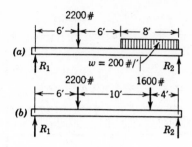

FIGURE 4-10

Taking R_1 as the center of moments,

$$18R_2 + (200 \times 4) = (1000 \times 8) + (800 \times 14) + (600 \times 20)$$
$$18R_2 = 30,400 \quad \text{and} \quad R_2 = 1688.9 \text{ lb}$$

Check:

$$200 + 1000 + 800 + 600 = 911.1 + 1688.9$$
$$2600 \text{ lb} = 2600 \text{ lb}$$

Example 3. The simple beam shown in Fig. 4-10a has one concentrated load of 2200 lb and a uniformly distributed load of 200 lb per lin ft extending over a distance of 8 ft. Compute the reactions.
Solution: Since the uniformly distributed load extends over a length of 8 ft, the total uniformly distributed load is 200×8 or 1600 lb. As noted earlier, such a load produces the same reactions as a concentrated load of the same magnitude applied at its center of gravity. As this load covers a distance of 8 ft, its center of gravity lies at a point 4 ft from the right end of the beam. Insofar as determination of reactions is concerned, this loading is the same as that shown in Fig. 4-10b. Returning to Fig. 4-10a, the equation of moments with R_2 as the center of moments is

$$20R_1 = (2200 \times 14) + (200 \times 8 \times 4)$$

In this equation, 4 is the moment arm of the distributed load about R_2:

$$20R_1 = 37,200 \quad \text{and} \quad R_1 = 1860 \text{ lb}$$

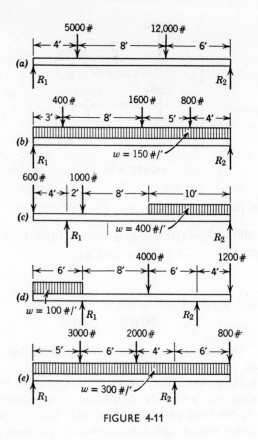

FIGURE 4-11

Taking R_1 as the center of moments,

$$20R_2 = (2200 \times 6) + (200 \times 8 \times 16)$$
$$20R_2 = 38,800 \quad \text{and} \quad R_2 = 1940 \text{ lb}$$

Check:

$$2200 + 1600 = 1860 + 1940$$
$$3800 \text{ lb} = 3800 \text{ lb}$$

Problems 4-6-A-B-C*-D-E*. Compute the reactions for the beams and loads shown in Fig. 4-11a, b, c, d, and e.

4-7 Vertical Shear

As explained in Art. 2-3, the *vertical shear* (*V*) is the tendency for one part of a member to move vertically with respect to an adjacent part. Referring to Fig. 2-3, if the total uniformly distributed load on the beam is 14,400 lb, the loading is symmetrical, and each reaction is 14,400/2 or 7200 lb. If we imagine a vertical plane cutting a section through the beam flush with the face of the left supporting wall, the reaction of the wall on the beam to the left of the section is upward. Just to the right of the section the load on the beam acts downward, and the magnitude of the tendency for the left and right portions to slide past each other (the vertical shear) is equal to the value of the reaction, 7200 lb. Thus we say that, at a section through the beam infinitely close to the support, *V* = 7200 lb.

At any other vertical section taken along the beam span, this same tendency for the portion to the left of the section to slip past the portion to the right exists. We may, therefore, formulate the following definition:

The magnitude of the vertical shear at any section of a beam is equal to the algebraic sum of all the vertical forces on one side of the section.

In order to understand how readily the value of the vertical shear may be computed for any section of a beam, let us call the upward forces (the reactions) positive and the downward forces (the loads) negative. For the present, let us consider only the forces to the *left* of the section. Then, in accordance with the foregoing definition, we can say *the magnitude of the vertical shear at any section of a beam is equal to the sum of the reactions minus the sum of the loads to the left of the section.* To find the value of the shear at a particular section, we simply note the forces to the left and then write: shear = reactions − loads. Of course, we could consider the forces to the right of the section and find that the magnitude of the shear will be the same, but in the following examples we shall consider the forces on the left. This procedure will avoid confusion with respect to algebraic sign, which will be discussed later.

Let us try this rule on the beam shown in Fig. 4-7. We have found that the left reaction is 450 lb and that 1350 lb is the magnitude of the right reaction. When writing an equation for computing the value of the vertical shear, it is convenient to indicate the section at which the shear is taken by using a subscript: $V_{(x=4)}$. For this illustration, this terminology indicates that the value of the vertical shear is taken at 4 ft from the left end of the beam.

Now referring to Fig. 4-7 and remembering that $R_1 = 450$ lb and $R_2 = 1350$ lb, let us write the values of the shear at 2 ft from R_1. Repeat the foregoing rule and observe that the only force to the left of the section is 450 lb, the reaction. Then

$$shear = reaction - loads$$

or

$$V_{(x=2)} = 450 \qquad -0$$

Thus,

$$V_{(x=2)} = 450$$

Since there are no loads up to the load of 1800 lb, the value of the shear is the same magnitude at any section in this portion of the beam. Note, however, that we are ignoring the relatively light weight of the beam.

Next consider a section 10 ft from R_1. The reaction to the left of this section is 450 lb, and the load to the left is 1800 lb. Then, since the shear equals reactions minus loads,

$$V_{(x=10)} = 450 - 1800 = -1350 \text{ lb}$$

Since this beam has but one load, the value of the shear at any section between the load and the right reaction is found to be -1350 lb. It should be noted here that for *simple* beams the value of the shear at the support is equal to the magnitude of the reaction. Hence the maximum shear will have the value of the greater reaction.

There is no physical significance to the difference between negative and positive shear. A widely accepted convention defines the shear at a section as positive $(+)$ if the portion of the beam *to the left* of the section tends to move upward with respect to the portion to the right; the shear is negative $(-)$ if the *right portion* tends to move upward with respect to the left portion. This convention enables one to work from either end of a beam and have the algebraic signs consistent.

4-8 Use of Shear Values

We compute the value of the shear in beams for two principal reasons. First, we must know the value of the maximum shear to be sure that there is ample material in the beam to prevent it from failing by shear. The second reason for computing the shear values is that *the greatest tendency for the beam to fail by bending is at the section of the beam at which the value of the shear is zero*. In the foregoing illustration, we found both positive and negative values for the shear, and we also found that values on the left of the 1800 lb load were positive and those on the right were negative. The critical section for bending is under the concentrated load where the shear changes from positive to negative. This will be explained further when we construct shear diagrams.

4-9 Shear Diagrams

In the preceding article, we saw that the value of the shear may be computed readily at any section along the beam's length. When designing beams, it will be found that diagrams showing the variation in shear along the span are extremely helpful.

To construct a shear diagram, first make a drawing of the beam to scale, showing the loads and their positions. Next, compute the reactions as explained in Art. 4-6. Below the beam, draw a horizontal *base line* representing zero shear. Then the values of the shear at various sections of the beam may be computed and plotted to a suitable scale vertically from the base line, positive values above and negative values below. The following several examples illustrate the construction of shear diagrams for beams under various conditions of loading.

Example 1. A simple beam 15 ft long has two concentrated loads located as shown in Fig. 4-12a. Construct the shear diagram and note the value of the maximum shear and the section of the beam at which the shear passes through zero.

Solution: Taking R_2 as the center of moments,

$$15R_1 = (8000 \times 12) + (12{,}000 \times 5)$$

$$15R_1 = 156{,}000 \quad \text{and} \quad R_1 = 10{,}400 \text{ lb}$$

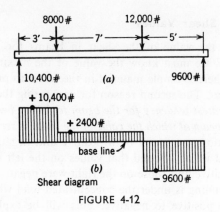

FIGURE 4-12

With R_1 as the center of moments,

$$15R_2 = (8000 \times 3) + (12{,}000 \times 10)$$
$$15R_2 = 144{,}000 \quad \text{and} \quad R_2 = 9600 \text{ lb}$$

Now let us compute the value of the shear at various sections.

$$V_{(x=1)} = 10{,}400 - 0 = 10{,}400 \text{ lb}$$
$$V_{(x=4)} = 10{,}400 - 8000 = 2400 \text{ lb}$$
$$V_{(x=11)} = 10{,}400 - (8000 + 12{,}000) = -9600 \text{ lb}$$

Since there are only two concentrated loads, it is unnecessary to compute the shear for other sections. For instance, the value of the shear at any section between the two loads is 2400 lb.

Now draw the base line shown in Fig. 4-12b. At a distance 1 ft to the right of R_1, lay off vertically 10,400 lb, at any convenient scale of so many pounds to the inch, *above* the line. Since the shear has this same value at any section between R_1 and the 8000-lb load, a series of points thus located would, when joined, result in a horizontal line.

Immediately to the right of the 8000 lb, for instance at $x = 4$, the value of the shear was found to be 2400 lb, still a *positive* quantity. This same value continues up to the 12,000-lb load and the points are plotted at the same scale *above* the base line because the values are positive.

At the section 11 ft from R_1, we have found the shear value to be −9600 lb, a *negative* quantity. This point is plotted *below* the line, as shown in the diagram. As the shear values are −9600 lb at all sections between the 12,000 lb load and R_2, a horizontal line is drawn through the point −9600 lb. This completes the shear diagram. The positive and negative values of the shear are hatched with vertical lines, as shown. Observe that the value of the shear at any section of the beam is the vertical distance given in the shear diagram.

The value of the maximum shear is seen to be 10,400 lb, the magnitude of the left reaction. On inspecting the shear diagram, we see that the shear passes through zero (the base line) at $x = 10$ ft, directly under the 12,000-lb load. Later on we shall see in the design of beams that it is very important to know the position of this section. It is here that the greatest bending stresses occur.

Thus far we have considered a beam with concentrated loads; now let us make a shear diagram for a uniformly distributed load.

Example 2. The beam shown in Fig. 4-13 has a span of 18 ft and a uniformly distributed load of 500 lb per lin ft. Construct the shear diagram for this beam and loading, and note the maximum shear and the section at which the shear passes through zero.
Solution: Since the beam is 18 ft in length and has a load of 500 lb per lin ft, the total uniformly distributed load is 500 × 18, or 9000 lb. The reactions are equal since the beam is symmetrically loaded, and $R_1 = R_2 = 9000/2 = 4500$ lb. The value of the shear

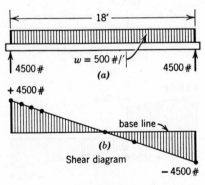

FIGURE 4-13

at a section immediately to the right of R_1 is 4500 − 0, or 4500 lb. Now let us compute the shear 1 ft to the right of R_1. $V_{(x=1)} = 4500 - 500 = 4000$ lb. Note that at this section the reaction to the left is 4500 lb and the load to the left is 500 lb, the load per linear foot. Several other sections are considered, and the values of the shear are computed as follows:

$$V_{(x=2)} = 4500 - (500 \times 2) = 3500 \text{ lb}$$
$$V_{(x=3)} = 4500 - (500 \times 3) = 3000 \text{ lb}$$
$$V_{(x=9)} = 4500 - (500 \times 9) = 0$$
$$V_{(x=12)} = 4500 - (500 \times 12) = -1500 \text{ lb}$$
$$V_{(x=18)} = 4500 - (500 \times 18) = -4500 \text{ lb}$$

The base line is drawn and the values of the shear at the various sections are plotted. A line drawn through the points is a sloping straight line. The shear has maximum values at the reactions, 4500 lb, and passes through zero at the center of the span.

Example 3. The simple beam shown in Fig. 4-14a has a span of 16 ft, a concentrated load of 6000 lb located 6 ft from the left support, and a uniformly distributed load of 200 lb per lin ft extending over the full length of the beam. Draw the shear diagram and note the maximum shear and the section at which the shear passes through zero.

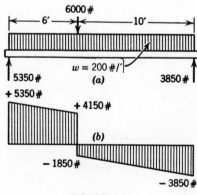

FIGURE 4-14

Solution: To compute the reactions, first take R_2 as the center of moments. Then

$$16R_1 = (6000 \times 10) + (200 \times 16 \times 8)$$
$$16R_1 = 85{,}600 \quad \text{and} \quad R_1 = 5350 \text{ lb}$$

With R_1 as the center of moments,

$$16R_2 = (6000 \times 6) + (200 \times 16 \times 8)$$
$$16R_2 = 61{,}600 \quad \text{and} \quad R_2 = 3850 \text{ lb}$$

As we have now computed the reactions, we know that the shear at R_1 is $+5350$ lb. Now let us compute the value of the shear at a section infinitely close to and to the left of the 6000-lb load. We call this distance $(x = 6-)$ from R_1. Then

$$V_{(x=6-)} = 5350 - (6 \times 200) = 4150 \text{ lb}$$

The next section is immediately to the right of the 6000 lb load; we can call it $(x = 6+)$ from R_1. Then

$$V_{(x=6+)} = 5350 - [(6 \times 200) + 6000] \quad = -1850 \text{ lb}$$
$$V_{(x=16)} = 5350 - [(16 \times 200) + 6000] = -3850 \text{ lb}$$

These various points are plotted and the shear diagram is drawn as shown in Fig. 4-14*b*. We find that the maximum shear is 5350 lb and that the shear passes through zero directly under the 6000-lb load.

Example 4.   Construct the shear diagram for the beam shown in Fig. 4-15*a*. Note the maximum shear and the section at which the shear passes through zero.

Solution: With R_2 as the center of moments,

$$20R_1 = (800 \times 10 \times 15) + (4000 \times 6)$$
$$20R_1 = 144{,}000 \quad \text{and} \quad R_1 = 7200 \text{ lb}$$

With R_1 as the center of moments,

$$20R_2 = (800 \times 10 \times 5) + (4000 \times 14)$$
$$20R_2 = 96{,}000 \quad \text{and} \quad R_2 = 4800 \text{ lb}$$

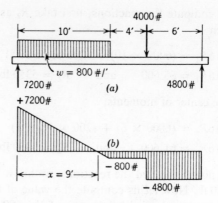

FIGURE 4-15

By computing the values of the shear, we find the shear at R_1 to be 7200 lb.

$$V_{(x=10)} \ \ = 7200 - (10 \times 800) = -800 \text{ lb}$$
$$V_{(x=14-)} = 7200 - (10 \times 800) = -800 \text{ lb}$$
$$V_{(x=14+)} = 7200 - [(10 \times 800) + 4000] = -4800 \text{ lb}$$
$$V_{(x=20)} \ \ = 7200 - [(10 \times 800) + 4000] = -4800 \text{ lb}$$

These various points are plotted and the shear diagram is drawn as shown in Fig. 4-15b. The value of the maximum shear is 7200 lb.

Now we find something new. Inspecting the shear diagram, we see that the shear passes through zero at some section between R_1 and the end of the distributed load. This point is at a distance of something less than 10 ft from R_1. What is its exact location?

This is a simple matter. We call the unknown distance x and write an equation for the shear at this section. The term x will occur in this equation, and the solution of the equation determines the location of zero shear. Then

$$V = 7200 - (800 \times x)$$

But we know that the value of the shear at this section is zero. Then

$$0 = 7200 - (800 \times x)$$
$$800x = 7200 \qquad \text{and} \qquad x = 9 \text{ ft}$$

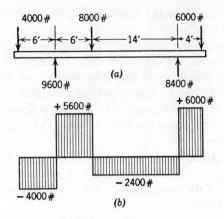

FIGURE 4-16

Since the shear passes through zero at a section of the beam 9 ft from R_1, this is the section at which the bending stresses will be the greatest.

Example 5. Construct the shear diagram for the overhanging beam and loads shown in Fig. 4-16a.

Solution: Writing an equation of moments with R_2 as the center of moments gives

$$20R_1 + (6000 \times 4) = (4000 \times 26) + (8000 \times 14)$$
$$20R_1 = 192{,}000 \quad \text{and} \quad R_1 = 9600 \text{ lb}$$

With R_1 as the center of moments,

$$20R_2 + (4000 \times 6) = (8000 \times 6) + (6000 \times 24)$$
$$20R_2 = 168{,}000 \quad \text{and} \quad R_2 = 8400 \text{ lb}$$

Then

$$V_{(x=1)} = -4000 \text{ lb}$$
$$V_{(x=7)} = 9600 - 4000 = 5600 \text{ lb}$$
$$V_{(x=13)} = 9600 - (4000 + 8000) = -2400 \text{ lb}$$
$$V_{(x=27)} = (9600 + 8400) - (4000 + 8000) = 6000 \text{ lb}$$

The shear at other sections might be computed, but we have sufficient information for our purpose. The values computed are now plotted with respect to the base line and the shear diagram shown in Fig. 4-16b results.

This diagram differs somewhat from those previously constructed. For simple beams, the maximum shear is the value of the greater reaction. But this is an overhanging beam, and we see that the maximum shear has a value of 6000 lb. We find also that in this beam the shear passes through zero at three different points, at the two supports and under the 8000-lb load. The significance of this will be explained later.

Problems 4-9-A-B-C-D*-E*-F. Construct the shear diagrams for the beams and loads shown in Fig. 4-17a, b, c, d, e, and f. In each case, note the magnitude of the maximum shear and the section at which the shear passes through zero.

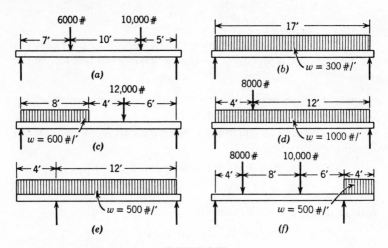

FIGURE 4-17

4-10 Bending Moments

As noted earlier, a beam deforms by bending under the action of applied loads. Figure 4-1a shows the deformation curve for a simple beam supporting a concentrated load at midspan; the deformation curve for the beam of Fig. 4-18a would be substantially similar.[3] At any point between the left reaction and the 4000-lb

---

[3] It should be noted that the deflections indicated by deformation curves are not normally visible to the eye but are nevertheless present.

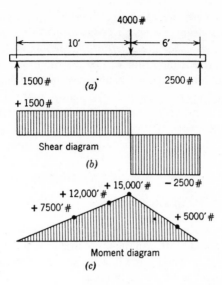

FIGURE 4-18

load, the tendency for the beam to bend is measured by the moment of the left reaction about the point in question. For example, consider a point 5 ft to the right of the left support. The moment of the reaction about this point as center of moments is 1500×5, or 7500 ft-lbs. This moment is called the *bending moment*, and its value may be plotted as a point on a moment diagram (Fig. 4-18*c*). The magnitude of the bending moment varies at different sections along the beam span; at 8 ft from R_1, for instance, it is 1500×8, or 12,000 ft-lb, and at 10 ft from R_1, under the 4000-lb load, it is 1500×10, or 15,000 ft-lb. Suppose, at this same section, we take the forces to the right instead of the forces to the left. Then the bending moment is 2500×6, or 15,000 ft-lb, a bending moment of the same magnitude.

The bending moment at any section of a beam is equal to the algebraic sum of the moments of all the forces on one side of the section.

For convenience, call the upward forces (the reactions) positive quantities, the downward forces (the loads) negative, and consider the forces to the left of the section. Then, in conformity with the foregoing statement, we can make the rule: *the bending moment at any section of a beam is equal to the moments of the reactions minus the moments of the loads to the left of the section.* If the forces to the right are considered, the result will be the same.

For those unfamiliar with the subject, it may be confusing at first to distinguish between shear and bending moments. Remember that the shear is *reactions minus loads*, whereas the bending moment is *moments of reactions minus moments of loads*. The shear is in units of pounds or kips, and the bending moment in units of ft-lb of kip-ft.

4-11 Bending Moment Diagrams

Figure 4-18c is a bending moment diagram, three ordinates of which were computed and plotted in the preceding article. Bending moment diagrams are constructed quite like the shear diagrams. A base line is drawn below the beam diagram (or shear diagram) and the values of the bending moments at various sections of the beams are plotted to scale. Using the subscripts $x = 1$, $x = 2$, etc. to indicate the section at which the bending moment is taken, we compute the magnitudes of the moments in accordance with the rule given in Art. 4-10. The section at which the moment is to be computed, as well as the reactions and loads to the *left* of this section, are noted. Then we write the moment of the reaction and subtract the moments of the loads. Applying this procedure to the beam in Fig. 4-18a, the bending moments at a few additional sections along the span (measured from the left end) are as follows:

$$M_{(x=0)} = 0$$
$$M_{(x=1)} = 1500 \times 1 = 1500 \text{ ft-lb}$$
$$M_{(x=2)} = 1500 \times 2 = 3000 \text{ ft-lb}$$
$$M_{(x=6)} = 1500 \times 6 = 9000 \text{ ft-lb}$$
$$M_{(x=10)} = 1500 \times 10 = 15{,}000 \text{ ft-lb}$$
$$M_{(x=12)} = (1500 \times 12) - (4000 \times 2) = 10{,}000 \text{ ft-lb}$$
$$M_{(x=14)} = (1500 \times 14) - (4000 \times 4) = 5000 \text{ ft-lb}$$
$$M_{(x=16)} = (1500 \times 16) - (4000 \times 6) = 0$$

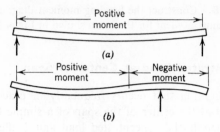

FIGURE 4-19

These values, being positive, are plotted above the base line, and the points are connected to form the diagram of Fig. 4-18c. In the design of beams, we are particularly concerned with the maximum value of the bending moment. For this beam, it is 15,000 ft-lb and occurs directly under the 4000-lb load. *Note that this is the section at which the shear passes through zero.* Attention is also called to the fact that for simple beams the value of the bending moment at the supports is zero.

4-12   Positive and Negative Bending Moments

Figure 4-19a shows the shape a simple beam tends to assume when it bends. The fibers in the upper surface of the beam are in compression, and those in the lower surface are in tension. We say the beam is "concave upward" and define the bending moment under this condition as *positive*. Now refer to Fig. 4-19b, an over-hanging beam. From the sketch, we see that *a portion* of the beam in the vicinity of the right reaction has tension stresses in the fibers of the upper surface and compression in the lower; here the beam is "concave downward," and the bending moment is *negative*. The section at which the bending moment changes from positive to negative is called the *inflection point*. We shall find later that at this section the bending moment has a value of zero. Negative bending moments are, of course, plotted below the base line in bending moment diagrams. (See Fig. 4-24).

Problem 4-12-A.* Construct the bending moment diagram for the beam shown in Fig. 4-12a and note the value of the maximum bending moment.

Problem 4-12-B. Construct the bending moment diagram and note the maximum bending moment for the beam shown in Fig. 4-8.

4-13 Concentrated Load at Center of Span

A loading condition that occurs frequently in practice is a concentrated load at the center of the span of a simple beam. In Fig. 4-20a, let P denote the concentrated load and L the span length. Then, as the beam is symmetrical, each reaction is equal to $P/2$ and the shear diagram is drawn as shown in Fig. 4-20b. From this diagram, it is seen that the shear passes through zero at the center of the span, at $x = L/2$. At this section, the value of the bending moment will be maximum. By following the rule given in Art 4-10, its magnitude is readily found as follows:

$$M_{(x=L/2)} = \frac{P}{2} \times \frac{L}{2} = \frac{PL}{4} \qquad \text{(See Fig. 4-20c.)}$$

This is a most useful expression. For instance, suppose we are to design a beam with a span of 24 ft and a concentrated load of 18 kips at midspan. It will be necessary to compute the maximum bending

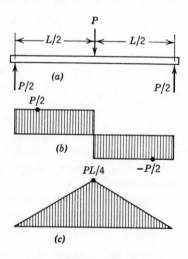

FIGURE 4-20

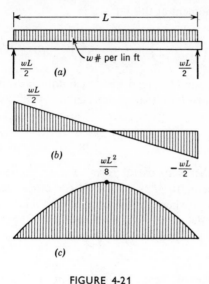

FIGURE 4-21

moment, which is readily accomplished for we have found its value to be $PL/4$, or

$$M = \frac{PL}{4} = \frac{18 \times 24}{4} = 108 \text{ kip-ft}$$

If the bending moment is required in units of kip-in., as it frequently is, $M = 108 \times 12 = 1296$ kip-in.

4-14 Simple Beam with Uniformly Distributed Load

Another condition that occurs perhaps more times than any other is a simple beam with a uniformly distributed load (Fig. 4-21a). To develop a formula for its maximum bending moment, let L be the span in feet and w the uniform load in lb per lin ft. Then the total distributed load is wL, each reaction will be $wL/2$, and the shear diagram will have the form shown in Fig. 4-21b. See also Fig. 4-13. The maximum bending moment is at the center of the span, at $x = L/2$. Then, by applying the rule given in Art. 4-10, the value of the bending moment at this section is

$$M = \left(\frac{wL}{2} \times \frac{L}{2}\right) - \left(\frac{wL}{2} \times \frac{L}{4}\right) \quad \text{or} \quad M = \frac{wL^2}{8}$$

In this expression, the load per linear foot is w lb, and the total load is wL. If the total load is represented by W, we have

$$W = wL \quad \text{and} \quad M = \frac{wL^2}{8} \quad \text{or} \quad M = \frac{WL}{8}$$

which is another very useful form of the equation for maximum bending moment. If M is computed for other sections along the span, the values when plotted form a parabola, as shown in Fig. 4-21c. The reader should satisfy himself that the bending moment at the quarter point of the span ($x = L/4$ from the left reaction) is $3wL^2/32$.

Example.   A simple beam having a span of 22 ft has a uniformly distributed load of 800 lb per lin ft. Compute the maximum bending moment.
Solution: Substituting in the formula for maximum bending moment for the given loading conditions,

$$M = \frac{wL^2}{8} = \frac{800 \times 22 \times 22}{8} = 48,400 \text{ ft-lb}$$

This is the value of the bending moment at the center of the span. It is computed quickly by using the formula. If, however, we were required to construct the entire bending moment diagram, it would be necessary to compute the values at several other sections so that a smooth curve could be plotted. In this example, R_1 is equal to half the total load on the beam or 8800 lb, and the bending moments at sections located 4, 8, 11, and 16 ft from R_1 are computed as follows:

$$M_{(x=4)} = (8800 \times 4) - (800 \times 4 \times 2) = 28,800 \text{ ft-lb}$$
$$M_{(x=8)} = (8800 \times 8) - (800 \times 8 \times 4) = 44,800 \text{ ft-lb}$$
$$M_{(x=11)} = (8800 \times 11) - (800 \times 11 \times 5.5) = 48,400 \text{ ft-lb}$$
$$M_{(x=16)} = (8800 \times 16) - (800 \times 16 \times 8) = 38,400 \text{ ft-lb}$$

When these values are plotted, they fall on a parabola similar to that shown in Fig. 4-21c.

Problem 4-14-A.   A simple beam has a span of 16 ft and a uniformly distributed load of 600 lb per lin ft. Compute the bending moment at the center and at the quarter points of the span, and draw the shear and bending moment diagrams.

4-15   Maximum Bending Moments

In order to compute the magnitude of the maximum bending moment for a beam, we must first know its position along the span. For an unsymmetrical loading, it is generally necessary to construct the shear diagram. Then, as noted earlier, the section at which the shear passes through zero is the section at which maximum bending moment occurs. The following examples relate to unsymmetrically loaded beams and show the method of determining the maximum bending moments.

Example 1.   The simple beam shown in Fig. 4-22a has three concentrated loads at the positions indicated. Compute the maximum bending moment.

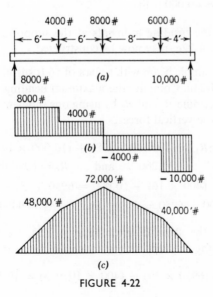

FIGURE 4-22

Solution: To find the reactions, we may determine R_1 by writing a moment equation and R_2 by summing the vertical forces.

$$24R_1 = (4000 \times 18) + (8000 \times 12) + (6000 \times 4)$$
$$24R_1 = 192{,}000 \quad \text{and} \quad R_1 = 8000 \text{ lb}$$
$$4000 + 8000 + 6000 = 8000 + R_2$$
$$18{,}000 - 8000 = R_2 \quad \text{and} \quad R_2 = 10{,}000 \text{ lb}$$

The value of the shear is computed at critical sections, and the shear diagram is constructed as shown in Fig. 4-22b. From this diagram, we find that the shear passes through zero under the 8000-lb load. The maximum bending moment is, therefore,

$$M_{(x=12)} = (8000 \times 12) - (4000 \times 6) = 72{,}000 \text{ ft-lb}$$

If we wish to construct the bending moment diagram, the expressions for moments under the other loads are

$$M_{(x=6)} = 8000 \times 6 = 48{,}000 \text{ ft-lb}$$
$$M_{(x=20)} = (8000 \times 20) - [(4000 \times 14) + (8000 \times 8)]$$
$$= 40{,}000 \text{ ft-lb}$$

Recalling that the bending moments at the supports are zero, the diagram of Fig. 4-22c, may now be constructed.

Example 2. A simple beam with a span of 16 ft sustains the loading shown in Fig. 4-23a. Compute the maximum bending moment.
Solution: Determine R_1 and R_2 by using one moment equation and then summing the vertical forces.

$$16R_1 = (400 \times 16 \times 8) + (10{,}000 \times 6)$$
$$16R_1 = 111{,}200 \quad \text{and} \quad R_1 = 6950 \text{ lb}$$
$$(400 \times 16) + 10{,}000 = 6950 + R_2$$
$$16{,}400 - 6950 = R_2 \quad \text{and} \quad R_2 = 9450 \text{ lb}$$

Constructing the shear diagram (Fig. 4-23b), we find that the maximum bending moment occurs under the concentrated load:

$$M_{(x=10)} = (6950 \times 10) - (400 \times 10 \times 5) = 49{,}500 \text{ ft-lb}$$

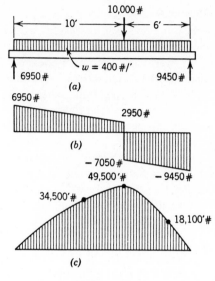

FIGURE 4-23

This is plotted to scale in Fig. 4-23c. If we wish to draw the bending moment diagram more accurately, two additional values may be computed as follows:

$$M_{(x=6)} = (6950 \times 6) - (400 \times 6 \times 3) = 34{,}500 \text{ ft-lb}$$
$$M_{(x=14)} = (6950 \times 14) - [(400 \times 14 \times 7) + (10{,}000 \times 4)]$$
$$= 18{,}100 \text{ ft-lb}$$

Example 3.   Compute the maximum bending moment for the beam shown in Fig. 4-15a.

Solution: Inspection of the loading diagram (Fig. 4-15a) gives no indication of the position of the maximum bending moment. However, after computing the reactions and drawing the shear diagram (Fig. 4-15b), it was found that the shear passed through zero at a section located 9 ft from the left support. The value of the maximum bending moment is, therefore,

$$M_{(x=9)} = (7200 \times 9) - (800 \times 9 \times 4.5) = 32{,}400 \text{ ft-lb}$$

4-16 Overhanging Beams

For the simple beams previously discussed, we have seen that the
maximum shear has the same magnitude as the greater reaction.
This is not true for overhanging beams. For example, we found that
the beam shown in Fig. 4-16, Art. 4-9, had R_1 as the greater reaction,
with a value of 9600 lb. The maximum shear, on the other hand,
was 6000 lb and occurred just to the right of R_2. We noted also that
the shear diagram for this beam passed through zero at three sections
along its length, so it is necessary to determine at which of these
sections the actual maximum bending moment occurs. This determination is accomplished by constructing the shear and moment
diagrams, exercising particular care when plotting the positive and
negative values. We will take as an example the same loading shown
in Fig. 4-16 but restated below as Fig. 4-24.

Example.   Construct the shear and bending moment diagrams for
the overcharging beam indicated in Fig. 4-24a, and note the value
of the maximum shear and maximum bending moment.

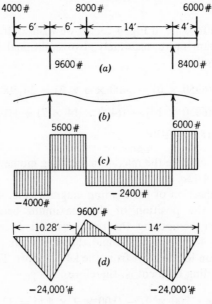

FIGURE 4-24

Solution: The shear diagram is constructed as explained for Fig. 4-16*b* and repeated here as Fig. 4-24*c*. The maximum shear is 6000 lb.

When determining bending moments in overhanging beams, it is very helpful to sketch the approximate deformation curve as indicated in Fig. 4-24*b*. (Compare Fig. 4-1*c* where the overhang occurs at one end only.) The curve for this beam and loading shows that we will encounter both positive and negative bending moments as defined in Art. 4-12. Now let us compute the bending moment values at certain sections.

$$M_{(x=6)} = -(4000 \times 6) = -24{,}000 \text{ ft-lb}$$

$$M_{(x=12)} = (9600 \times 6) - (4000 \times 12) = +9600 \text{ ft-lb}$$

$$M_{(x=26)} = (9600 \times 20) - [(4000 \times 26) + (8000 \times 14)]$$

$$= -24{,}000 \text{ ft-lb}$$

Other values might be computed; but since there are no uniformly distributed loads, it is unnecessary. The values just computed are plotted with respect to the base line, positive values above and negative values below; the completed moment diagram is shown in Fig. 4-24*d*. The points at which the bending moment diagram passes through zero are of special significance; they are discussed in the following article where their locating dimensions are also determined. The maximum value of the bending moment is seen to be 24,000 ft-lb; it is negative, and it occurs over both supports. (It is only coincidental that the moment has identical values over the two supports.)

Where beams have complicated loading patterns, it is advantageous to compute bending moments from both ends in order to simplify the arithmetic. Where this is done, a procedure which will always give the correct sign for the bending moment is to consider the moments of all upward forces as positive $(+)$ and the moments of all downward forces as negative $(-)$. Using this convention, the bending moment over the right reaction in Fig. 4-24 may be computed as follows:

$$M = -6000 \times 4 = -24{,}000 \text{ ft-lb}$$

which is the same value as that obtained above, working from the left, for M at $x = 26$ ft.

4-17   Inflection Point

The two points in Fig. 4-24d where the bending moment diagram passes through zero are called the *inflection points* or *points of contraflexure*. The inflection point is the point or section along the beam length at which the curvature of the beam changes from "concave upward" to "concave downward" (Art. 4-12 and Fig. 4-19) and at which the value of the bending moment is zero. Its position may be scaled from the moment diagram or computed by letting x be the distance from the end of the beam to the inflection point and writing an expression for the bending moment at this section.

Example 1.   Determine the position of the inflection point to the right of the left reaction for the beam shown in Fig. 4-24.
Solution: Let x be the distance from the inflection point to the left end of the beam. The expression for bending moment at this point is

$$M = [9600 \times (x - 6)] - (4000 \times x)$$

But we know that the value of the moment at this section is zero. Then

$$0 = [9600 \times (x - 6)] - (4000 \times x)$$
$$9600x - 57,600 = 4000x$$
$$5600x = 57,600 \quad \text{and} \quad x = 10.28 \text{ ft}$$

To find the position of the inflection point to the left of the right support, it will simplify the mathematics if we consider the moments of the forces to the *right* of the section instead of the left. Let x be the distance from the inflection point to the right end of the beam. Then, as before,

$$0 = [8400 \times (x - 4)] - (6000 \times x)$$
$$8400x - 33,600 = 6000x$$
$$2400x = 33,600 \quad \text{and} \quad x = 14 \text{ ft}$$

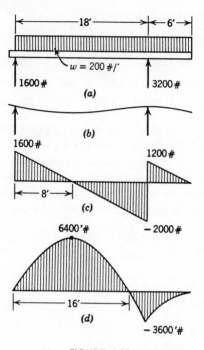

FIGURE 4-25

Example 2. The overhanging beam shown in Fig. 4-25*a* has a uniformly distributed load of 200 lb per lin ft over its entire length. Construct the shear and moment diagrams, note the values of maximum shear and maximum bending moment, and compute the position of the inflection point.

Solution: The curve the bent beam will take is approximated in Fig. 4-25*b*; we shall expect to find both positive and negative bending moments. Computing the reactions,

$$18R_1 = 200 \times 24 \times 6$$
$$18R_1 = 28{,}800 \qquad \text{and} \qquad R_1 = 1600 \text{ lb}$$
$$18R_2 = 200 \times 24 \times 12$$
$$18R_2 = 57{,}600 \qquad \text{and} \qquad R_2 = 3200 \text{ lb}$$

Computing the values of the shear,

$$V \text{ (at left support)} = 1600 - 0 = 1600 \text{ lb}$$
$$V_{(x=18-)} = 1600 - (200 \times 18) = -2000 \text{ lb}$$
$$V_{(x=18+)} = (1600 + 3200) - (200 \times 18) = 1200 \text{ lb}$$
$$V_{(x=24)} = (1600 + 3200) - (200 \times 24) = 0$$

The shear diagram is plotted as shown in Fig. 4-25c, and the maximum shear value is 2000 lb. Since the shear passes through zero at two points, the bending moment diagram will have maximum values at two places; one will be a positive and the other a negative moment. The maximum negative moment is directly above the right support. To find the position of zero shear between the supports,

$$0 = 1600 - (200 \times x)$$
$$200x = 1600 \quad \text{and} \quad x = 8 \text{ ft}$$

The bending moment will have values of zero at each end of the beam. Values for the two critical sections are computed thus:

$$M_{(x=8)} = (1600 \times 8) - (200 \times 8 \times 4) = 6400 \text{ ft-lb}$$
$$M_{(x=18)} = (1600 \times 18) - (200 \times 18 \times 9) = -3600 \text{ ft-lb}$$

The maximum bending moment is 6400 ft-lb, the positive moment.

To find the position of the inflection point, let x be its distance from the left support. The value of the bending moment at this section is zero. Then

$$(1600 \times x) - \left(200 \times x \times \frac{x}{2}\right) = 0$$

$$\frac{200x^2}{2} - 1600x = 0$$

$$100x^2 - 1600x = 0$$

$$x^2 - 16x = 0$$

Completing the square,

$$x^2 - 16x + 64 = 64$$

Extracting the square root of both sides,

$$x - 8 = 8 \quad \text{and} \quad x = 16\,\text{ft}$$

4-18   Cantilever Beams

The computations for shear and bending moments for the portion of cantilever beams extending beyond the face of the supporting wall are quite simple. The reaction at the wall is equal to the sum of the loads on the beam. If the beam diagram is drawn with the support at the right, the rules given in Arts. 4-7 and 4-10 for computing shear and moments are used. The maximum values for both the shear and bending moment are at the face of the wall.

Example.   The cantilever beam shown in Fig. 4-26a has a concentrated load at the free end and a uniformly distributed load applied over the first 6 ft of length from the wall. Construct the shear and bending moment diagrams, and note the maximum values of each.

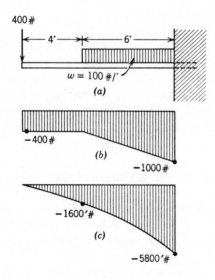

FIGURE 4-26

Solution: The value of the shear is computed at various sections:

$$V_{(x=1)} = -400\,\text{lb}$$
$$V_{(x=4)} = -400\,\text{lb}$$
$$V_{(x=6)} = -[400 + (2 \times 100)] = -600\,\text{lb}$$
$$V_{(x=8)} = -[400 + (4 \times 100)] = -800\,\text{lb}$$
$$V_{(x=10)} = -[400 + (6 \times 100)] = -1000\,\text{lb}$$

The maximum value of the shear is 1000 lb, and it occurs at the face of the wall. The shear diagram is plotted as shown in Fig. 4-26*b*.

Computing the values of the bending moments at various sections,

$$M_{(x=1)} = -(400 \times 1) = -400\,\text{ft-lb}$$
$$M_{(x=2)} = -(400 \times 2) = -800\,\text{ft-lb}$$
$$M_{(x=4)} = -(400 \times 4) = -1600\,\text{ft-lb}$$
$$M_{(x=8)} = -[(400 \times 8) + (100 \times 4 \times 2)] = -4000\,\text{ft-lb}$$
$$M_{(x=10)} = -[(400 \times 10) + (100 \times 6 \times 3)] = -5800\,\text{ft-lb}$$

The maximum bending moment is at the face of the wall; its value is −5800 ft-lb. The curve of the bending moment diagram is a straight line up to the −1600 ft-lb value and is a portion of a parabola from there to the wall. In accordance with the convention established in Art. 4-12, the full length of a cantilever will be under negative bending moment.

4-19 Typical Loads for Simple and Cantilever Beams

A simple beam with the load at the center of the span and a simple beam with a uniformly distributed load over its entire length are conditions that occur frequently in practice. These and other common types of loading are shown in Fig. 4-27. This figure shows at a glance the values of the maximum shear, maximum bending moment, and maximum deflection (to be discussed later) for the

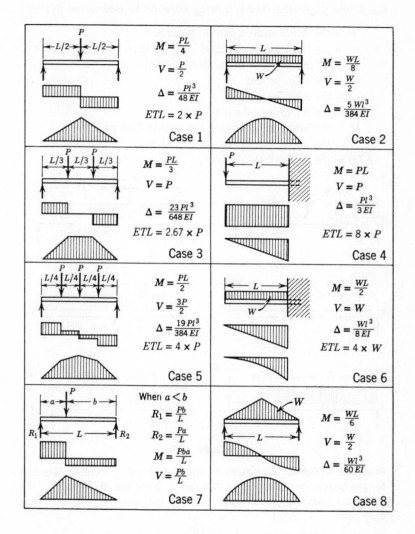

Case 1
$$M = \frac{PL}{4}$$
$$V = \frac{P}{2}$$
$$\Delta = \frac{Pl^3}{48\,EI}$$
$$ETL = 2 \times P$$

Case 2
$$M = \frac{WL}{8}$$
$$V = \frac{W}{2}$$
$$\Delta = \frac{5\,Wl^3}{384\,EI}$$

Case 3
$$M = \frac{PL}{3}$$
$$V = P$$
$$\Delta = \frac{23\,Pl^3}{648\,EI}$$
$$ETL = 2.67 \times P$$

Case 4
$$M = PL$$
$$V = P$$
$$\Delta = \frac{Pl^3}{3\,EI}$$
$$ETL = 8 \times P$$

Case 5
$$M = \frac{PL}{2}$$
$$V = \frac{3P}{2}$$
$$\Delta = \frac{19\,Pl^3}{384\,EI}$$
$$ETL = 4 \times P$$

Case 6
$$M = \frac{WL}{2}$$
$$V = W$$
$$\Delta = \frac{Wl^3}{8\,EI}$$
$$ETL = 4 \times W$$

Case 7
When $a < b$
$$R_1 = \frac{Pb}{L}$$
$$R_2 = \frac{Pa}{L}$$
$$M = \frac{Pba}{L}$$
$$V = \frac{Pb}{L}$$

Case 8
$$M = \frac{WL}{6}$$
$$V = \frac{W}{2}$$
$$\Delta = \frac{Wl^3}{60\,EI}$$

FIGURE 4-27

conditions presented. The following notation is used in the figure:

P = the concentrated load in pounds
W = the total uniformly distributed load in pounds (in Case 8, W represents a triangular loading)
L = the span in feet
l = the span in inches
V = the maximum shear in pounds
M = the maximum bending moment in foot-pounds
Δ = the maximum deflection in inches
ETL = the equivalent tabular load (see Art. 8-9)

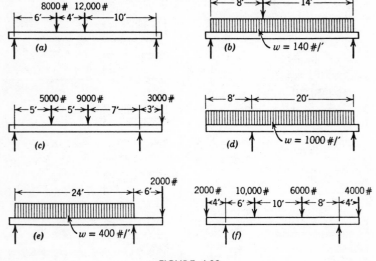

FIGURE 4-28

Problems 4-19-A-B. Construct the shear and bending moment diagrams for the beams shown in Fig. 4-28a and b. Note the magnitude of the maximum shear and maximum bending moments on the respective diagrams.

Problems 4-19-C*-D*-E-F. For the beams shown in Fig. 4-28c, d, e, and f, draw the shear and bending moment diagrams. For each beam, give the value of the maximum shear and maximum bending moment, and compute the position of the inflection points. Verify the location of the inflection points by measuring on the bending moment diagrams.

5

Theory of Bending and Properties of Sections

||||||||||||||||||||||||||||||||||||||||||||||||||

5-1   Resisting Moment

In the preceding chapter, the concept of bending moment was developed as a means of measuring the tendency of the external loads on a beam to deform it by bending. It is now necessary to consider the action within the beam that resists bending and is called the *resisting moment*.

A rectangular beam subjected to bending is shown in Fig. 5-1a. If we consider any section along the beam, such as X-X, the external forces produce a bending moment at this section. Letting x be the distance of the section from the left reaction, as indicated in the enlarged detail of Fig. 5-1b, the value of the bending moment is $R_1 \times x$. This bending moment is resisted by internal stresses set up in the fibers of the beam.

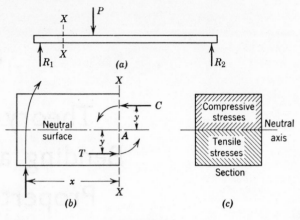

FIGURE 5-1

As pointed out in Art. 2-4, the stresses above the neutral surface are compressive and those below are tensile. Let C be the resultant of all the compressive stresses, T the resultant of all the tensile stresses, and y their distances from the neutral axis, point A. Then $(C \times y) + (T \times y)$ equals the sum of the moments of all the internal stresses at section X-X. Referring to Fig. 5-1b, we note that the moments of the stresses in the fibers tend to cause a counterclockwise rotation about point A, while the bending moment generated by R_1 tends to produce a clockwise rotation about the same point. The sum of the moments of all the internal stresses about the neutral axis is called the *resisting moment* because it holds the bending moment in equilibrium.

At any section along the length of a beam, the bending moment and the resisting moment are equal in magnitude; and using the foregoing terms, we may write

$$R_1 \times x = (C \times y) + (T \times y)$$

or

$$\text{bending moment} = \text{resisting moment}$$

The design of a beam consists primarily of (1) computing the maximum bending moment that will be developed by the design loading, and (2) selecting a structural steel shape that will provide a potential resisting moment equal to or greater than the maximum bending moment. The study of Chapter 4 has enabled us to compute the maximum bending moment for most of the steel beams that will be encountered in practice; the second part of the above procedure is accomplished by use of the *flexure formula* (frequently called the *beam formula*) developed in the following article.

5-2 The Flexure Formula

The flexure formula is an expression for resisting moment that involves the size and shape of the beam cross section and the material of which the beam is made. It is used in the design of all homogeneous beams (i.e., beams made of one material only such as steel, aluminum, or wood).

Figure 5-2 represents a portion of the side elevation and the cross section of a homogeneous beam subject to bending stresses. The cross section shown is unsymmetrical about the neutral axis, but this discussion applies to a cross section of any shape. In Fig. 5-2a, let c be the distance of the most remote fiber from the neutral axis, and let f be the unit stress on the fiber at c distance. If f, the greatest fiber stress, does not exceed the elastic limit of the material, the stresses in the other fibers are directly proportional to their distances from the neutral axis. If c is in inches, the unit stress on a fiber at 1 in. distance is f/c. To explain this, suppose $c = 8$ in. and the stress, f, on the most remote fiber is 1600 psi. Then the stress on a fiber at 1 in. from the neutral axis is f/c, or 1600/8, or 200 psi; at 2 in., it is 2×200, or 400 psi; at 3 in., it is 3×200, or 600 psi, etc.

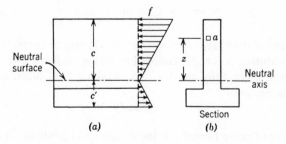

(a)   Section (b)

FIGURE 5-2

Now imagine an infinitesimally small area at z distance from the neutral axis and call this very small area a. Since the unit stress on a fiber at unity distance is f/c, the unit stress on the fiber at z distance must be $f/c \times z$. Since there are a square inches in this elementary area, the stress on the fiber a will be $f/c \times z \times a$. The *moment* of the stress on fiber a about the neutral axis will be the stress times its lever arm, or

$$\left(\frac{f}{c} \times z \times a\right) \times z \quad \text{or} \quad \frac{f}{c} \times a \times z^2$$

We know there is an infinite number of these elementary areas in the cross section; and if we use the symbol $\sum$ to indicate the sum of an infinite number, $\sum \frac{f}{c} \times a \times z^2$ will represent the sum of the moments of all the stresses in the cross section (both above and below the neutral axis) with respect to the neutral axis. But this sum, we know from Art. 5-1, is the *resisting moment*. Since we know that the bending moment and resisting moment at any section of a beam under load are equal in magnitude, we may write

$$M = M_R \quad \text{or} \quad M = \frac{f}{c}\sum a z^2$$

The term $\sum az^2$ is called the moment of inertia of the cross section with respect to the neutral axis, and it is discussed in more detail in Art. 5-5. It is usually designated by the letter I. Using this symbol, the above expression may be written

$$M = \frac{f}{c} \times I \quad \text{or} \quad M = \frac{fI}{c}$$

and is called the *beam* or *flexure formula*. It may be simplified further by substituting S for I/c, called the *section modulus*, a term that is described more fully in Art. 5-6. Making this substitution, the formula becomes

$$M = fS$$

Use of the flexure formula in the design and investigation of beams is explained in Chapter 6.

5-3   Properties of Sections

Each of the terms, moment of inertia and section modulus, used in deriving the flexure formula represents a *property* of a particular beam cross section. Other properties are the area of the section, the radius of gyration, and the position of the centroid of the cross-sectional area. Although these properties and other useful design constants are tabulated for the commonly used structural shapes (Tables 1-1 through 1-5), a knowledge of their significance will be of great value to the designer. Several of the properties listed in the tables are discussed below, and others are considered in Chapter 8 under the design of beams.

5-4   Centroids

The term *centroid* is related to plane areas in the same manner that *center of gravity* is related to solids. *The centroid of a plane surface is a point that corresponds to the center of gravity of a very thin homogeneous plate having the same area and shape.* It can be shown that the neutral axis of a beam cross section passes through its centroid; consequently, it is necessary to know its exact position. For symmetrical sections such as wide flange shapes and I-beams, it can be seen by inspection that the centroid lies at a point midway between the flanges, at the intersection of the X-X and Y-Y axes shown at the head of Tables 1-1 and 1-2. The distance c from the extreme fiber to the neutral axis is, therefore, half the depth of such sections. This is also true for standard channels used as beams when bending takes place about the X-X axis, although the position of the centroid with respect to its distance from the back of the web cannot be determined by inspection and must be computed. This situation is shown in the sketch at the head of Table 1-3, and computed values of $\bar{x}$ (read x bar) are given in the table. For example, the centroid of a C 10 × 20 is located 5 in. from the backs of the flanges and 0.606 in. from the back of the web (the distance $\bar{x}$).

For unsymmetrical sections such as angles, the centroid may be found by referring to Tables 1-4 and 1-5. As an illustration, consider an L 6 × 4 × ½. Table 1-5 shows that the centroid is 1.99 in. from the back of the short leg and 0.987 in. from the back of the long leg. In the sketch at the head of the table, these distances are indicated as y and x, respectively. As Table 1-4 shows, the x and y distances are equal for equal leg angles.

The procedure for computing the position of centroids utilizes the principle of *statical moments*. A statical moment of a plane area with respect to an axis is the area multiplied by the perpendicular distance from the centroid to the axis. If a plane area is divided into a number of parts, the sum of the statical moments of the parts is equal to the statical moment of the entire area. The following example illustrates how readily this principle is applied.

Example.   Compute the distance of the centroid from the back of the short leg of an L 6 × 4 × ½.

Solution: A diagram showing the dimensions of the angle is first made (Fig. 5-3); it is divided into the two areas indicated by the diagonals and marked A and B, but any convenient division will serve our purpose. The value sought is denoted c'. Area A is 6 × 0.5, or 3 sq in., and area B is 3.5 × 0.5, or 1.75 sq in. The area of the entire cross section is 3 + 1.75, or 4.75 sq in. Now, *with respect to an axis passing through the back of the short leg*, we may write the sum of the statical moments of the two parts and equate it to the statical moment of the entire area.

Noting that the distances of the centroids of rectangular areas A and B to the axis selected are 3 and 0.25 in., respectively, the equation becomes

$$(3 \times 3) + (1.75 \times 0.25) = 4.75 \times c'$$
$$9.438 = 4.75 \times c' \quad \text{and} \quad c' = 1.99 \text{ in.}$$

This is the same value given as y for this angle in Table 1-5.

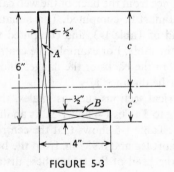

FIGURE 5-3

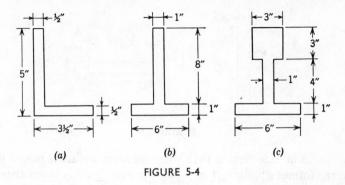

FIGURE 5-4

It should be observed that, if this angle were used as a beam with the long leg in the vertical position, the neutral axis would be located 1.99 in. above the bottom of the cross section. The distance of the *most remote fiber* from the neutral axis, however, would be $6 - 1.99$, or 4.01 in. This would be the distance c in the flexure formula.

Problems 5-4-A*-B-C.* Compute the values of c with respect to the horizontal axes of the beam cross sections shown in Fig. 5-4a, b, and c.

5-5 Moment of Inertia

When developing the flexure formula (Art. 5-2), we called the term $\sum az^2$ the moment of inertia of the beam cross section and denoted it by the symbol I. Moment of inertia may be defined as *the sum of the products obtained by multiplying all the elementary areas of a cross section by the squares of their distances from a given axis.* It may be found for any axis, but the axis commonly used is the neutral axis of a beam cross section, which, as we know, passes through the centroid of the section. Since the elements of area are expressed in square inches and the distances in inches, the units for moment of inertia will be inches raised to the fourth power, written *inches*[4].

In order to understand more clearly the significance of moment of inertia, we may observe from the expression $\sum az^2$ that I varies with the *shape* of the cross section as well as with the area. It is, of course, a matter of common observation that a wood plank, say 2 in. by 6 in. in cross section, is much stiffer when used on a given span

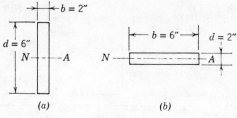

FIGURE 5-5

with its 6-in. side vertical than when the same surface is placed flat. In the former situation (Fig. 5-5a), the term $\sum az^2$ is larger than in the latter (Fig. 5-5b) since the average value of z is larger. It follows, therefore, that it is desirable to place as much material as far as practicable from the neutral axis when seeking to maximize the value of I.[1]

The moment of inertia of any section may be computed by use of the calculus or other summation method. However, I values for structural steel shapes are listed in tables giving properties of structural sections. As an example, refer to Table 1-1 which gives data relating to wide flange shapes. Under W 18 × 45 we find that the I of this section with respect to the X-X axis is 706 in.[4] If we also look up the moment of inertia of a W 16 × 45 (a shape having the same weight per linear foot and cross-sectional area), we find its $I_{X\text{-}X}$ to be 584 in.[4] Comparing these two values, the effect of the 2-in. difference in depth on the moments of inertia is readily apparent. As an aid to fixing in mind the concept of moment of inertia, it is recommended that the reader make similar comparisons between other sections listed in the tables.

5-6 Section Modulus

As noted in Art. 5-2, the term I/c in the flexure formula is called the *section modulus*. It is defined as the moment of inertia divided by the distance of the most remote fiber from the neutral axis and is

---

[1] The formula for moment of inertia of a rectangle about an axis through its centroid and parallel to its base is $I = bd^3/12$ (see Table 5-1). Referring to Fig. 5-5, the reader should satisfy himself that the value of I with the longer side vertical is 36 in.[4], while with the longer side flat it is 4 in.[4]

denoted by the symbol S. Since I and c always have the same values for any given cross section, values of S may be computed and tabulated for structural shapes. With I expressed in inches to the fourth power and c a linear dimension in inches, S is in units of inches to the third power, written *inches*.[3] Table 1-1 gives values of the section modulus with respect to the two major axes of wide flange shapes. For example, a W 12 × 40 has an $S_{X\text{-}X}$ of 51.9 in.[3] and an $S_{Y\text{-}Y}$

TABLE 5-1   Properties of Geometric Sections

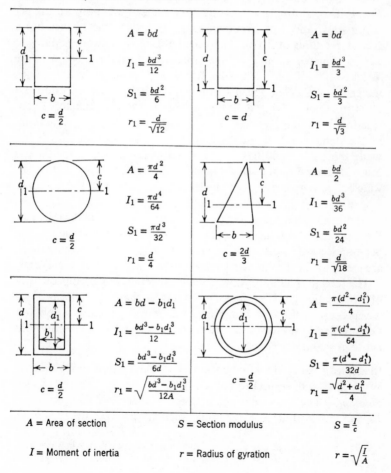

Rectangle (axis through centroid):
$$A = bd$$
$$I_1 = \frac{bd^3}{12}$$
$$S_1 = \frac{bd^2}{6}$$
$$c = \frac{d}{2} \qquad r_1 = \frac{d}{\sqrt{12}}$$

Rectangle (axis at base):
$$A = bd$$
$$I_1 = \frac{bd^3}{3}$$
$$S_1 = \frac{bd^2}{3}$$
$$c = d \qquad r_1 = \frac{d}{\sqrt{3}}$$

Circle:
$$A = \frac{\pi d^2}{4}$$
$$I_1 = \frac{\pi d^4}{64}$$
$$S_1 = \frac{\pi d^3}{32}$$
$$c = \frac{d}{2} \qquad r_1 = \frac{d}{4}$$

Triangle:
$$A = \frac{bd}{2}$$
$$I_1 = \frac{bd^3}{36}$$
$$S_1 = \frac{bd^2}{24}$$
$$c = \frac{2d}{3} \qquad r_1 = \frac{d}{\sqrt{18}}$$

Hollow rectangle:
$$A = bd - b_1 d_1$$
$$I_1 = \frac{bd^3 - b_1 d_1^3}{12}$$
$$S_1 = \frac{bd^3 - b_1 d_1^3}{6d}$$
$$c = \frac{d}{2} \qquad r_1 = \sqrt{\frac{bd^3 - b_1 d_1^3}{12A}}$$

Hollow circle:
$$A = \frac{\pi(d^2 - d_1^2)}{4}$$
$$I_1 = \frac{\pi(d^4 - d_1^4)}{64}$$
$$S_1 = \frac{\pi(d^4 - d_1^4)}{32d}$$
$$c = \frac{d}{2} \qquad r_1 = \frac{\sqrt{d^2 + d_1^2}}{4}$$

A = Area of section S = Section modulus $S = \frac{I}{c}$

I = Moment of inertia r = Radius of gyration $r = \sqrt{\frac{I}{A}}$

of 11.0 in.[3] We will be concerned mostly with S_{X-X} since structural shapes are normally used as beams with the X-X axis as the axis about which bending takes place.

Like moment of inertia, the section modulus is related to the size and shape of the beam cross section, and shapes having a high percentage of material placed in the flanges will have larger values of S than other configurations of the same cross-sectional area. This will be apparent from a comparison of the wide flange shape W 24 × 100 (Table 1-1) and the Standard I-beam S 24 × 100 (Table 1-2). Both these sections have the same depth, weight per linear foot, and cross-sectional area.[2] However, the former has a section modulus of 250 in.,[3] and the latter has a section modulus of 199 in.[3] Further scanning of the tables reveals that the flange of the W 24 × 100 is 12 in. wide and 0.775 in. thick, making the area of one flange 9.3 sq in. The flange of the S 24 × 100 is 7.247 in. wide and 0.871 in. thick, giving a flange area of 6.3 sq in. Therefore, considering both flanges, the W 24 × 100 has 2 × 9.3 = 18.6 sq in. in the flanges out of the total cross-sectional area of 29.5 sq in., while the S 24 × 100 has only 2 × 6.3 = 12.6 sq in. in the flanges out of the same total area. As noted in Art. 1-2, this means that the wide flange shapes are more efficient structurally than the Standard I-beams.

Another reason for placing as much material as practicable near the top and bottom faces of beams is that the maximum value of the bending stress (f_b) occurs there (see Fig. 2-4d). These considerations are the basis for the familiar I and H shapes of structural steel beams.

The section modulus is one of the most commonly used properties of cross sections; it is employed constantly in the design of beams and is a measure of the ability to resist bending stresses. The example and two problems that follow should serve to fix in mind the relationship between it and moment of inertia.

Example.   Verify the tabulated value of the section modulus of a W 18 × 45 with respect to the X-X axis.

Solution: Table 1-1 gives the moment of inertia of this section as 706 in.[4] Since it is a symmetrical section, $c = d/2$ or $17.86/2 =$

---

[2] Within computational tolerance of the tables. The discrepancy between 29.5 sq in. listed for W 24 × 100 and 29.4 sq in. for S 24 × 100 may be disregarded.

8.93 in. Therefore,

$$S = \frac{I}{c} = \frac{706}{8.93} = 79.0 \text{ in.}^3$$

which checks with the value given in the table.

Problem 5-6-A. Verify the value of S_{X-X} given in Table 1-3 for the American Standard Channel C 15 × 50.

Problem 5-6-B. Verify the value of S_{X-X} given in Table 1-5 for the unequal leg angle L 5 × 3 × ½.

5-7 Radius of Gyration

This property of a cross section is related to the design of compression members rather than beams and will be discussed in more detail under the design of columns in Chapter 10. Since it is listed in the tables of properties, however, it will be considered briefly here.

Just as the section modulus is a measure of the resistance of a beam section to bending, the radius of gyration (which is also related to the size and shape of the cross section) is an index of the stiffness of a structural section when used as a column or other compression member. The radius of gyration is found by the formula

$$r = \sqrt{\frac{I}{A}}$$

and is expressed in inches since the moment of inertia is in inches[4] and the cross-sectional area is in square inches. For W, S, and C shapes (Tables 1-1, 1-2, and 1-3), r is tabulated for both the X-X and Y-Y axes; and for angles, an additional value is given with respect to an oblique axis through the centroid marked Z-Z. When used in the design of columns, the *least* radius of gyration is the one generally used.

5-8 Properties of Built-Up Sections

When rolled steel shapes are combined to form built-up structural sections such as those illustrated in Fig. 5-6, their properties for use in design must be computed. The first step in such computations is

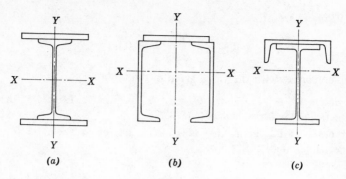

FIGURE 5-6

to determine the moment of inertia of the built-up section about its neutral axis. This usually requires transferring the moments of inertia of the cross sections of the individual parts from one axis to another since the moment of inertia of the built-up section about its neutral axis is equal to the sum of the moments of inertia of the individual parts about the same axis.

The simple built-up section shown in Fig. 5-7 consists of an S 10 × 25.4 with 12-in. by ½-in. plates welded to its flanges. Since the X-X axis of the S 10 × 25.4 coincides with the neutral axis NA of the built-up section, its I with respect to NA may be found from Table 1-2 directly and is 124 in.⁴ The moment of inertia of one of the plates about an axis through its centroid and parallel to its longer side (called its *gravity axis* and marked *ga* in Fig. 5-7) may be

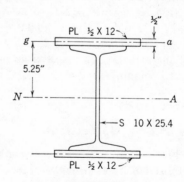

FIGURE 5-7

found from the formula for rectangular cross sections. This is

$$I = \frac{bd^3}{12} = \frac{12 \times 0.5^3}{12} = 0.125 \text{ in.}^4$$

In order to find the I of the plate about the NA of the built-up section, it is necessary to use the *transfer formula*, which may be stated as follows:

The moment of inertia of a cross section about any axis parallel to an axis through its own centroid is equal to the moment of inertia of the cross section about its own gravity axis plus its area times the square of the distance between the two axes.

Expressed mathematically,

$$I = I_o + Az^2$$

In this formula,

$I =$ the moment of inertia of the cross section about the required axis.

$I_o =$ the moment of inertia of the cross section about its own gravity axis parallel to the required axis.

$A =$ the area of the cross section.

$z =$ the distance between the two parallel axes.

In order to apply this equation to the *two* plates shown in Fig. 5-7, we note that the area of each plate is $12 \times 0.5 = 6$ sq in. and that the distance between ga and NA is 5.25 in. Then

$$\begin{aligned} I &= 2[I_o + Az^2] \\ &= 2[0.125 + (6 \times 5.25 \times 5.25)] \\ &= 2[0.125 + 165.375] = 2 \times 165.5 = 331 \text{ in.}^4 \end{aligned}$$

The moment of inertia of the entire built-up section about the neutral axis NA is, then, $124 + 331 = 455$ in.^{4}

Since this built-up section is symmetrical with respect to NA, its section modulus will be equal to I divided by half the overall depth, or

$$S = \frac{I}{c} = \frac{455}{5.5} = 82.7 \text{ in.}^3$$

5-9 Unsymmetrical Built-Up Sections

In the preceding article, the location of the neutral axis of the built-up section shown in Fig. 5-7 was determined by inspection. This was possible because the cross section is symmetrical about *NA*, and consequently its centroid lies at the midpoint of the overall depth. When working with unsymmetrical sections, however, it is necessary to locate the centroid by utilizing the principle of statical moments discussed in Art. 5-4. The application of this principle to unsymmetrical built-up sections is explained in the following example.

Example.   The built-up section shown in Fig. 5-8 is to be used as a lintel to carry a masonry wall over an opening. It consists of a C 12 × 20.7 to which an L 4 × 3 × $\frac{5}{16}$ has been welded with the longer leg outstanding. Determine its section modulus with respect to its *X-X* axis.

Solution: The first step is to locate the *X-X* axis. This is accomplished by finding the distance of the centroid from either the upper or lower flange of the channel. We will use the upper flange as our reference surface.

By referring to Table 1-3, we find the area of a C 12 × 20.7 to be 6.09 sq in. Since the channel's depth is 12 in., its centroid is 12/2 or

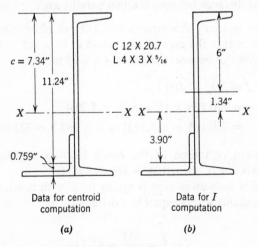

Data for centroid computation	Data for *I* computation
(a)	(b)

FIGURE 5-8

6 in. from the back of the top flange. Table 1-5 shows the area of an L 4 × 3 × $\frac{5}{16}$ to be 2.09 sq in.; it also shows that its centroid is 0.759 in. from the back of the long leg. This dimension is recorded on Fig. 5-8a, together with the distance of the angle's centroid from the upper flange of the channel (12 − 0.759 = 11.24 in.).

The combined area of the channel and angle is 6.09 + 2.09 = 8.18 sq in.

Writing an equation for statical moments about the *upper flange of the channel as an axis* and letting c be the distance of the centroid of the entire section from the back of the upper flange, we obtain the following equation:

$$8.18 \times c = (6.09 \times 6) + (2.09 \times 11.24)$$
$$8.18 \times c = 60.03 \quad \text{and} \quad c = 7.34 \text{ in.}$$

This dimension is recorded in Fig. 5-8a.

Having found the value of c, we have determined the position of the neutral axis of the entire cross section, and the next step is to transfer the moments of inertia of the channel and angle to the axis X-X.

The moment of inertia of the channel about an axis through its centroid, parallel to the flanges, is found from Table 1-3 to be 129 in.[4] The distance from this gravity axis to axis X-X of the built-up section is 7.34 − 6 or 1.34 in. as recorded in Fig. 5-8b. Then

$$I = I_o + Az^2$$
$$= 129 + (6.09 \times 1.34 \times 1.34)$$
$$= 129 + 10.9 = 139.9 \text{ in.}[4]$$

which is the moment of inertia of the channel with respect to the X-X axis.

Now consider the angle. Table 1-5 shows that its moment of inertia about an axis through its centroid parallel to its long leg is 1.65 in.[4] The distance of this gravity axis from axis X-X of the built-up section is 11.24 − 7.34 or 3.90 in., also recorded in Fig. 5-8b. Again,

$$I = I_o + Az^2$$
$$= 1.65 + (2.09 \times 3.90 \times 3.90)$$
$$= 1.65 + 31.8 = 33.5 \text{ in.}[4]$$

which is the moment of inertia of the angle with respect to the X-X axis.

Having computed the moments of inertia of both the channel and the angle with respect to the neutral axis, X-X, the moment of inertia of the entire section is equal to their sum, or

$$I_{X\text{-}X} = 139.9 + 33.5 = 173.4 \text{ in.}^4$$

The section modulus of the built-up section is, then

$$S = \frac{I}{c} = \frac{173.4}{7.34} = 23.6 \text{ in.}^3$$

Review Problems

Problem 5-9-A.* The built-up section shown in Fig. 5-6c consists of a W 14 × 30 with a C 10 × 15.3 welded to the top flange. Determine the moment of inertia and the section modulus with respect to the X-X axis.

Problem 5-9-B. Compute the value of I and S with respect to the Y-Y axis of the built-up section described in the preceding problem.

Problem 5-9-C.* For the angle section shown in Fig. 5-4a, compute the moment of inertia, section modulus, and radius of gyration with respect to an axis through the centroid parallel to the short leg.

Problem 5-9-D. An S 12 × 31.8 has an 8-in. by ½-in. plate welded to its upper flange. Compute the moment of inertia and section modulus for this built-up section with respect to an axis through its centroid parallel to the plate.

Problem 5-9-E.* A built-up section similar in shape to that shown in Fig. 5-8 is composed of a C 15 × 33.9 and an L 6 × 4 × ⅜ with the long leg outstanding. Compute the moment of inertia and the section modulus about an axis through its centroid perpendicular to the web of the channel.

Problem 5-9-F. A built-up section similar in shape to that shown in Fig. 5-6c is composed of a W 16 × 36 and a C 12 × 20.7 Compute the moment of inertia and the section modulus with respect to an axis through its centroid perpendicular to the web of the wide flange beam.

6

Use of
the Beam
Formula

||||||||||||||||||||||||||||||||||||||||||||||||||||

6-1 Forms of the Equation

The expression $M = fI/c$ or fS, developed in Art. 5-2, may be stated in three different forms depending upon the information sought. These are given below using the AISC general nomenclature, which makes a distinction between allowable bending stress (F_b) and computed bending stress (f_b).

$$\text{(1)}\ M = \frac{F_b I}{c} \qquad \text{(2)}\ f_b = \frac{Mc}{I} \qquad \text{(3)}\ \frac{I}{c} = \frac{M}{F_b}$$

Or letting $I/c = S$,

$$\text{(1)}\ M = F_b S \qquad \text{(2)}\ f_b = \frac{M}{S} \qquad \text{(3)}\ S = \frac{M}{F_b}$$

Form (1) gives the maximum potential resisting moment when the section modulus of the beam and the maximum allowable bending stress are known. Form (2) gives the computed bending stress when the maximum bending moment due to the loading is known,

101

together with the section modulus of the beam. These are the two forms used in investigation.

Form (3) is the one used in design. It gives the *required* section modulus when the maximum bending moment and the allowable bending stress are known. When the required section modulus has been determined, a beam having an S equal to or greater than the computed value is selected from tables giving properties of the various structural shapes.

When using the beam formula, care must be exercised with respect to the units in which the terms are expressed. Bending stress values F_b and f_b may be written in pounds per square inch (psi) or kips per square inch (ksi); S is stated in inches[3] and I in inches.[4] M, therefore, must be in inch-pounds or kip-inches. M, as customarily computed from the loads and reactions, is expressed in foot-pounds or kip-feet and must be converted to inch-pounds or kip-inches by multiplying its value by 12 before it is used in the formula.

6-2 Investigation of Beams

The process of determining whether a beam of given size and span can safely support a proposed loading is called investigation. Use of the beam formula in making such determinations is illustrated in the following examples.

Example 1. It is proposed to use a W 10 × 25 to carry a total uniformly distributed load of 30 kips (including an allowance for its own weight) on a span of 13 ft (Fig. 6-1). If the allowable bending stress is 24 ksi, determine whether the beam is safe (a) by comparing the maximum resisting moment of the section with the maximum bending moment developed by the loading; and (b) by comparing

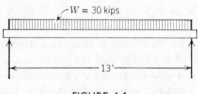

FIGURE 6-1

the allowable bending stress with that actually produced by the loading.

Solution (a):   Referring to Table 1-1, W Shapes-Properties for designing, the section modulus for the W 10 × 25 is found to be 26.5 in.[3] Then

$$M = F_b S = 24 \times 26.5 = 636 \text{ kip-in.}$$

$$M = 636/12 = 53.0 \text{ kip-ft}$$

From Art. 4-14 and Fig. 4-27 (Case 2), we know that the maximum bending moment for the proposed loading occurs at midspan and may be found from the formula $M = WL/8$. Then

$$M = \frac{WL}{8} = \frac{30 \times 13}{8} = 48.8 \text{ kip-ft}$$

The beam is safe as long as the developed bending moment (48.8 kip-ft) is less than the permissible resisting moment (53.0 kip-ft) determined above.

Solution (b):   The maximum bending stress will occur at the top and bottom surfaces (of this symmetrical section) where the largest bending moment occurs. Then

$$f_b = \frac{M}{S} = \frac{48.8 \times 12}{26.5} = 22.1 \text{ ksi}$$

This also verifies that the beam is safe since the actual extreme fiber stress (22.1 ksi) is less than the allowable (24.0 ksi). Note that the bending moment in the above equation was multiplied by 12 to convert it to kip-in.

Example 2.   An S 12 × 31.8 has a span of 14 ft. If the allowable bending stress is 22,000 psi, find the maximum concentrated load it will support at midspan (Fig. 6-2).

Solution: Referring to Table 1-2, S Shapes-Properties for designing, the section modulus for this I-beam is found to be 36.4 in.[3] The maximum resisting moment of the beam is

$$M = F_b S = 22,000 \times 36.4 = 801,000 \text{ in-lb}$$

$$M = 801,000/12 = 66,750 \text{ ft-lb}$$

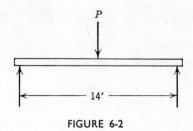

FIGURE 6-2

From Art. 4-13 and Fig. 4-27 (Case 1), we find that the maximum bending moment for this loading occurs at midspan and is given by the formula $M = PL/4$. However, before solving this equation for P, the bending moment due to the beam weight must be deducted from the maximum resisting moment. The moment due to beam weight also occurs at midspan and is found from the expression $M = wL^2/8$, as shown in Art. 4-14. Then

$$\dot{M} = \frac{wL^2}{.8} = \frac{3.18 \times 14 \times 14}{8} = 779 \text{ ft-lb}$$

and the resisting moment available to support the proposed loading is

$$66{,}750 - 779 = 65{,}971 \qquad \text{or} \qquad 65{,}970 \text{ ft-lb}$$

Therefore, the maximum safe concentrated load at the center of the span is

$$\frac{PL}{4} = M \qquad \text{or} \qquad P = \frac{4M}{L} = \frac{4 \times 65{,}970}{14} = 18{,}900 \text{ lb}$$

Example 3. For the unsymmetrical built-up section shown in Fig. 5-8, determine (a) its maximum resisting moment if the allowable bending stress is 24 ksi, and (b) the value of the bending stress at the bottom face of the section when the maximum resisting moment is being developed.

Solution (a): In the example of Art. 5-9, the section modulus of this built-up shape was found to be 23.6 in.[3] Therefore, its maximum resisting moment is

$$M = F_b S = 24 \times 23.6 = 566 \text{ kip-in.}$$

Solution (*b*): The computed value of *c* (distance of most remote fiber from neutral axis) is recorded on Fig. 5-8a as 7.34 in. Since the overall depth of the section is 12 in., the distance from the neutral axis to the bottom face is

$$c' = d - c = 12 - 7.34 = 4.66 \text{ in.}$$

Also, the value of *I* was determined to be 173.4 in.[4] Substituting these values in form (2) of the beam formula (Art. 6-1), the bending stress at the bottom face of the section is

$$f_b = \frac{Mc}{I} = \frac{Mc'}{I} = \frac{566 \times 4.66}{173.4} = 15.3 \text{ ksi}$$

It should be noted that this value is considerably less than the allowable 24 ksi which occurs simultaneously at the upper surface of the channel. In other words, the steel below the neutral axis is not being used to its full capacity. This is a characteristic of unsymmetrical sections, and although their use is sometimes required by special construction situations, they are not economical for general use as beams.

Problem 6-2-A.* An S 10 × 25.4 has a span of 10 ft with a uniformly distributed load of 36 kips in addition to its own weight. The allowable bending stress is 24 ksi. Is the beam safe with respect to bending stresses?

Problem 6-2-B. A W 16 × 45 has a loading consisting of 10 kips each at the quarter points of a 24-ft span and a uniformly distributed load of 5.2 kips including the beam weight. If the allowable bending stress is 24 ksi, is the beam safe with respect to bending stresses?

Problem 6-2-C.* Two 5 × 3½ × 1½-in. angles fastened together back-to-back are to be used as a beam on a span of 5 ft. The allowable bending stress is 22 ksi. Find the total permissible uniformly distributed load (a) when the long legs are placed vertically back-to-back and (b) when the short legs are so placed.

6-3 Design of Beams for Bending

As noted in Art. 6-1, form (3) of the beam formula is the one used in design. It gives the required section modulus after the maximum bending moment has been computed and the allowable bending stress determined from the specifications. The complete design of a beam includes consideration of additional items such as deflection, lateral support of the compression flange against buckling, shear, and web crippling. These will be dealt with in succeeding chapters, but the cardinal principle underlying beam design is that the beam cross section shall have a potential resisting moment equal to or greater than the bending moment. The beam formula provides the mechanism for establishing this basic condition.

Example.   Design a simply supported beam (Art. 4-2) to carry a superimposed[1] load of 2 kips per ft over a span of 24 ft. A36 steel (Art. 3-1 and Table 3-2) is to be used with an allowable bending stress of 24 ksi.

Solution: (1) The first of the steps in this solution (which are numbered for ready reference) is to make a sketch of the beam showing the loads, as in Fig. 6-3.

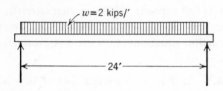

FIGURE 6-3

(2) The bending moment due to the superimposed load is

$$M = \frac{wL^2}{8} = \frac{2 \times 24 \times 24}{8} = 144 \text{ kip-ft}$$

(3) The required section modulus for this bending moment is

$$S = \frac{M}{F_b} = \frac{144 \times 12}{24} = 72.0 \text{ in.}^3$$

(4) Scanning Table 1-1, a W 16 × 45 is found to have a section modulus of 72.5 in.[3] However, this value is so close to the 72.0 in.[3] required by the superimposed load that practically no margin is provided for the effect of the beam weight. Further scanning of Table 1-1 reveals that a W 16 × 50 has an S of 80.8 in.[3] and that a

---

[1] The term *superimposed load* will be used to denote any load other than the weight of a structural member itself.

W 18 × 45 has an S of 79.0 in.³ In the absence of any restriction on beam depth, the lightest weight section that will carry the load should be used. Therefore, try the W 18 × 45.

(5) The bending moment at the center of the span due to beam weight is

$$M = \frac{wL^2}{8} = \frac{45 \times 24 \times 24}{8} = 3240 \text{ ft-lb} \quad \text{or} \quad 3.24 \text{ kip-ft}$$

(6) The total bending moment at midspan is

$$M = 144 + 3.24 = 147.24 \quad \text{or} \quad 147 \text{ kip-ft}$$

(7) The section modulus required for this moment is

$$S = \frac{M}{F_b} = \frac{147 \times 12}{24} = 73.5 \text{ in.}^3$$

Inasmuch as this required value is less than the 79.0 in.³ of the W 18 × 45, this section is adopted.

6-4   Structural Design Methods

There are two different methods employed in designing steel beams for bending stresses. The first, called *allowable stress design*, is used in this chapter and in the major portion of the book. It is the more conventional procedure based on the idea of using F_b, an allowable extreme fiber stress, as a certain fraction of the yield stress (Art. 3-5). Such allowable stresses fall below the elastic limit of the material, and we speak of them as conforming with the elastic behavior of the material. This approach to the design of steel members has been standard practice for many years.

The second method, known as *plastic design*, is a more recent development and is based on the idea of computing an ultimate load and on utilizing a portion of the reserve strength, after initial yield stress has been reached, as part of the factor of safety (Art. 2-8). This method was introduced into the AISC Specification in 1963 and forms Part 2 of the 1969 Specification. A brief explanation of plastic design theory is given in Chapter 14.

Note: The following problems involve design for bending stresses only. A36 steel with an allowable bending stress of 24 ksi is to be used. Full lateral support (as mentioned in Art. 6-3) is to be assumed.

Problem 6-4-A. A simple beam has a span of 12 ft with a uniformly distributed load, including its own weight, of 28 kips. What size wide flange section should be used?

Problem 6-4-B.* Two concentrated loads of 20 kips each are placed at the third point of a simple beam with a span of 24 ft. What should be the size of the beam?

Problem 6-4-C.* A simple beam has a span of 14 ft with a concentrated load of 15 kips applied at midspan. In addition, there is a superimposed load of 1 kip per ft extending over the entire span. Design the beam.

7

Deflection
of Beams

IIIIIIIIIIIIIIIIIIIIIIIIIIIIIIIIIIIIIIIIIIIIIIIIII

7-1 Deflection

The deformation that accompanies the bending of a beam is called
deflection. If the beam is in a horizontal position, it is the vertical
distance moved from a horizontal line. The deflection of a beam may
not be apparent visually but it is, nevertheless, always present.
Figure 7-1 illustrates the deflection of a simple beam; we are principally concerned with its maximum value, which in this instance
occurs at the center of the span.

A beam may be strong enough to withstand the bending stresses
without failure, but the curvature may be so great that cracks appear
in suspended plaster ceilings, water collects in low spots on roofs,
and the general lack of stiffness results in an excessively springy floor.
When designing floor construction for buildings in which machinery
will be used, particular care should be given to the deflection of
beams since excessive deflection may be the cause of inordinate
vibrations or misalignments. Another fault that results from excessive deflection occurs where floor beams frame into girders. At these
points, there is a tendency for cracks to develop in the flooring
directly over the girder. This is illustrated in Fig. 7-2.

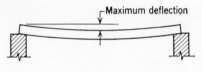

FIGURE 7-1

If, in the design of a beam, the deflection is computed to be excessive, the remedy is to select a deeper beam—a beam having a greater moment of inertia. For a given span and loading, the deflection of a beam varies directly as the fiber stress (bending stress) and inversely as the depth. For this reason, it is preferable to select a beam which has the greatest practicable depth so that the deflection will be a minimum. When ample headroom is available, a deeper beam is preferable to a shallow beam with the same section modulus.

In general, the procedure is to design a beam of ample dimensions to resist bending stresses (design for strength) and then to investigate the beam for deflection.

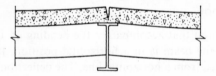

FIGURE 7-2

7-2 Allowable Deflection

There is common agreement that deflection of beams should be limited, but authorities differ with respect to the maximum degree of deflection that should be permitted. Some codes require that deflection be limited to $\frac{1}{360}$ of the span but say nothing of the kind of load that produces the deflection. The AISC Specification requires that beams and girders supporting plastered ceilings be of such dimensions that the maximum live load[1] deflection will not exceed

---

[1] The live load in a building is the probable load due to occupancy, as distinguished from dead load, which represents the weight of the construction. These terms are defined in more detail in Chapter 9, Floor Framing Systems.

⅓₆₀ of the span. In actual design work, the local building code should always be consulted as to specific provisions governing deflection.

7-3 Deflection for Uniformly Distributed Loads

Referring to Fig. 4-27, we find formulas that can be used to compute the maximum deflection for the several types of loading that occur most frequently in practice. The notation used in the formulas is given in Art. 4-19, and attention is called particularly to the lower case letter *l*, which denotes the length of the span in *inches*. The Greek letter delta (Δ) represents the maximum deflection; it, too, is in units of inches.

The slide rule is usually used to solve the equations for deflections. There are a number of terms in these formulas, and their solution may be a tedious task without the help of a slide rule. There are, however, simplified methods of computing deflections. Consider the following example.

A designer is engaged in determining the size of steel floor beams. He is using an allowable extreme fiber stress of 24,000 psi. His immediate problem is a simple beam with a span of 16 ft 0 in. and a uniformly distributed load of 34,000 lb. His computations show that a W 12 × 27 is large enough to resist the bending stresses, and his next step is to compute the maximum deflection.

To do this, he uses the very handy formula

$$\Delta = \frac{0.02483\ L^2}{d}$$

in which L = the span length of the beam in feet,
  d = the depth of the beam in inches,
and Δ = the maximum deflection of the beam in inches.

Referring to Table 1-1, we see that d for a W 12 × 27 is 11.96 in. Then

$$\Delta = \frac{0.02483 \times 16 \times 16}{11.96} \quad \text{and} \quad \Delta = 0.53\ \text{in.}$$

is the actual maximum deflection.

If the allowable deflection is $\frac{1}{360}$ of the span, $(16 \times 12)/360 = 0.53$ in. is the allowable deflection. The actual deflection *does not* exceed the allowable.

Now let us see how the foregoing formula, identified later as Fr. 7-3-3, is determined.

The maximum bending moment for this type of loading is $M = wl^2/8$. In Art. 6-1, we find

$$f = \frac{Mc}{I}$$

Consequently, substituting the value of M,

$$f = \frac{wl^2c}{8I}$$

The maximum deflection for a simple beam with a uniformly distributed load is given in Case 2, Fig. 4-27.

$$\Delta = \frac{5}{384} \times \frac{Wl^3}{EI}$$

or, since $W = wl$,

$$\Delta = \frac{5}{384} \times \frac{wl^4}{EI} = \left(\frac{wl^2c}{8I}\right)\left(\frac{5l^2}{48Ec}\right)$$

This may be written

$$\Delta = \frac{5}{48}\left(\frac{f}{E}\right)\left(\frac{l^2}{c}\right)$$

Since $l = 12L$,

$$\Delta = \frac{5}{48}\left(\frac{f}{E}\right)\left(\frac{144L^2}{c}\right)$$

and

$$\Delta = 15\left(\frac{f}{E}\right)\left(\frac{L^2}{c}\right) \qquad \text{(Formula 7-3-1)}$$

This is a basic formula; it can be used for any material by substituting the appropriate values for f and E.

For sections symmetrical with respect to a horizontal axis through the centroid, such as rectangular and I-sections, $c = d/2$. Then, substituting this value of c in Fr. 7-3-1,

$$\Delta = 15\left(\frac{f}{E}\right)\left(\frac{2L^2}{d}\right) \qquad \text{(Formula 7-3-2)}$$

Now let $f = 24{,}000$ psi and $E = 29{,}000{,}000$ psi. Substituting in Fr. 7-3-2,

$$\Delta = 15\left(\frac{24{,}000}{29{,}000{,}000}\right)\left(\frac{2L^2}{d}\right)$$

Thus, if $f = 24{,}000$ psi,

$$\Delta = \frac{0.02483L^2}{d} \qquad \text{(Formula 7-3-3)}$$

If $f = 22{,}000$ psi

$$\Delta = \frac{0.02276L^2}{d} \qquad \text{(Formula 7-3-4)}$$

If $f = 20{,}000$ psi,

$$\Delta = \frac{0.02069L^2}{d} \qquad \text{(Formula 7-3-5)}$$

Remember that these convenient formulas are to be used only for simple steel beams with uniformly distributed loads when the extreme fiber stress is approximately 24,000 psi, 22,000 psi, or 20,000 psi, as noted above.

If f (or f_b in AISC nomenclature), the extreme fiber stress developed by the loading, is unknown, we can use the deflection formulas given in Fig. 4-27.

Example 1. A simple beam has a span of 20 ft and a uniformly distributed load of 39,000 lb. The section used for this load is a W 14 × 34, and the extreme fiber stress is 24,000 psi. Compute the deflection.

Solution: Formula 7-3-3 is appropriate and, referring to Table 1-1, $d = 14$ in. Then

$$\Delta = \frac{0.02483L^2}{d} \qquad \text{or} \qquad \Delta = \frac{0.02483 \times 20 \times 20}{14} = 0.71 \text{ in.}$$

If we had not known that $f = 24{,}000$ psi, we might have used the formula for deflection given in Fig. 4-27, Case 2:

$$\Delta = \frac{5}{384} \times \frac{Wl^3}{EI}$$

Table 1-1 shows that I for a W 14×34 is 340 in.[4] Then

$$\Delta = \frac{5}{384} \times \frac{39{,}000 \times (20 \times 12)^3}{29{,}000{,}000 \times 340} \quad \text{and} \quad \Delta = 0.71 \text{ in.}$$

Note that this is the same value obtained by using Fr. 7-3-3.

Example 2.   A W 12×27 is used as a simple beam on a span of 19 ft. It supports a uniformly distributed load of 24,000 lb. If the extreme fiber stress is 20,000 psi, find the deflection.
Solution: Since in this problem $f = 20{,}000$ psi, Fr. 7-3-5 is appropriate. Referring to Table 1-1, we find the depth of a W 12×27 to be 11.96 in. Then

$$\Delta = \frac{0.02069L^2}{d} = \frac{0.02069 \times 19 \times 19}{11.96} = 0.62 \text{ in.}$$

7-4 Concentrated Loads

The deflections discussed in the preceding article concerned simple beams with uniformly distributed loads. The deflection for beams with concentrated loads may be computed by using the deflection formulas given in Fig. 4-27 or by using coefficients which relate the concentrated load deflection to that produced by a uniformly distributed load.

Formula 7-3-2 is the basic equation for computing the deflection due to distributed loads. The number 15 in the formula is the coefficient, usually denoted by C. In Table 7-1, coefficients of 12, 15.3, and 14.3 are given for Cases 1, 3, and 5, respectively, for beams with concentrated loads. Also shown in the table are ratios of the concentrated load deflections to $C = 15$.

Example 1.   A simple beam has a span of 16 ft with a concentrated load of 32,000 lb at the center of the span. The section used for this loading is a W 16×40, and the bending stress developed is 24,000 psi. Compute the deflection.

TABLE 7-1. Deflection Coefficients for Simple Beams

$$\Delta = C \left(\frac{f}{E}\right)\left(\frac{2L^2}{d}\right)$$

Cases Fig. 4-27	Type of loading	Loading diagram	Δ Deflection	C Deflection coefficient	Ratio of deflection to coefficient 15*
Case 1	Concentrated load at center of span		$\Delta = \frac{1}{48} \times \frac{Pl^3}{EI}$	12	0.80
Case 2	Uniformly distributed load		$\Delta = \frac{5}{384} \times \frac{Wl^3}{EI}$	15	1.00
Case 3	Concentrated loads at third points		$\Delta = \frac{23}{648} \times \frac{Pl^3}{EI}$	15.3	1.02
Case 5	Concentrated loads at fourth points		$\Delta = \frac{19}{384} \times \frac{Pl^3}{EI}$	14.3	0.95

* For irregular loading, the ratio of deflection to coefficient 15 is 0.92.

Solution: If this had been a uniformly distributed load, the deflection could have been determined by Fr. 7-3-3,

$$\Delta = \frac{0.02483L^2}{d}$$

But Table 7-1 shows that the deflection for a concentrated load at the center of the span is only 0.80 times this value. Then

$$\Delta = 0.80 \times \frac{0.02483 \times 16 \times 16}{16} = 0.32 \text{ in.}$$

If we had not known the magnitude of the bending stress, we could have used the equation for Case 1, Fig. 4-27, which is repeated in Table 7-1. Referring to Table 1-1, the moment of inertia for the W 16 × 40 is 517 in.[4] Then

$$\Delta = \frac{1}{48} \times \frac{Pl^3}{EI} \quad \text{or} \quad \Delta = \frac{1}{48} \times \frac{32,000 \times (16 \times 12)^3}{29,000,000 \times 517} = 0.32 \text{ in.}$$

Example 2. Three concentrated loads of 4000 lb each are applied at the quarter points of a 20-ft span. In addition, there is a uniformly distributed load of 20,000 lb extending over the entire beam length. The section used is a W 16 × 36. Compute the deflection.

Solution: Because the maximum deflection from the concentrated loading and that from the distributed load occur at the same point, center of span, the maximum deflection of the beam will be the sum of these deflections computed separately. Consider first the deflection that results from the concentrated loading. From Case 5, Fig. 4-27, we find

$$\Delta = \frac{19}{384} \times \frac{Pl^3}{EI}$$

and, referring to Table 1-1, I for the W 16 × 36 is given as 447 in.[4] Then

$$\Delta = \frac{19}{384} \times \frac{4000 \times (20 \times 12)^3}{29,000,000 \times 447} \quad \text{and} \quad \Delta = 0.21 \text{ in.}$$

which is the maximum deflection caused by the concentrated loads.

Next let us compute the deflection due to the uniformly distributed load. Case 2, Fig. 4-27, gives

$$\Delta = \frac{5}{384} \times \frac{Wl^3}{EI}$$

and substituting the known values

$$\Delta = \frac{5}{384} \times \frac{20,000 \times (20 \times 12)^3}{29,000,000 \times 447} \quad \text{and} \quad \Delta = 0.28 \text{ in.}$$

which is the maximum deflection that results from the uniformly distributed load.

Adding the deflections caused by the two types of loading, $0.21 + 0.28 = 0.49$ in., the maximum deflection of the beam. If the allowable deflection is $\frac{1}{360}$ of the span, the permissible value would be $(20 \times 12)/360 = 0.66$ in. The computed deflection is well within this limit.

Problem 7-4-A. A W 14 × 34 is used for a simple beam on a span of 18 ft. There is a uniformly distributed load of 43,000 lb extending over the full length of the beam, and the bending stress on the extreme fiber is 24,000 psi. Compute the deflection produced by this load.

Problem 7-4-B. A simple beam has a span length of 20 ft with a distributed load of 44 kips over its entire length. If the shape used is an S

15 × 42.9 and the bending stress is 22 ksi, determine the deflection caused by this superimposed load.

Problem 7-4-C.* A W 18 × 50 is used on a span of 24 ft. The loading consists of two concentrated loads of 22 kips, each located at the third points of the span. If the bending stress is 24 ksi, compute the deflection due to the concentrated loads.

Problem 7-4-D. An S 10 × 25.4 is used to support a uniformly distributed load, including its own weight, of 18 kips on a span of 15 ft. Compute the maximum deflection.

Problem 7-4-E.* A simple beam has a span of 24 ft with two concentrated loads of 40 kips each at the third points of the span and a uniformly distributed load of 24 kips extending over the entire length of the beam. The section used is a W 16 × 45. Compute the deflection due to this superimposed loading.

7-5   A Convenient Deflection Formula

Minute accuracy in computing the deflection of beams is seldom necessary. The designer usually wishes to know only whether the deflection is excessive. As we have seen, authorities are not in complete agreement concerning the allowable deflection. It is also well to bear in mind that the live loads used in computing deflections are *estimated* loads rather than quantities known with a high degree of precision.

We have seen that, for symmetrical shapes such as I-beams and wide flange sections, the deflection may be computed by the formula

$$\Delta = 15\left(\frac{f}{E}\right)\left(\frac{2L^2}{d}\right) \qquad (7\text{-}3\text{-}2)$$

The modulus of elasticity E of structural steel is 29,000,000 psi (Art. 2-7). Frequently, however, designers use 30,000,000 psi to simplify approximate computations and express both the bending stress and E in ksi. Formula 7-3-2 may then be written

$$\Delta = 15\left(\frac{f}{30,000}\right)\left(\frac{2L^2}{d}\right)$$

and

$$\Delta = \left(\frac{f}{1000}\right)\left(\frac{L^2}{d}\right) \qquad (\text{Formula } 7\text{-}5\text{-}1)$$

Formula 7-3-2 in this form is of great convenience. The answer, not exactly correct, is usually adequate for practical purposes.

Example. A simple beam has a span of 16 ft with a uniformly distributed load of 50 kips extending over its full length. The section used is an S 12 × 50 and the bending stress is 24 ksi. Compute the maximum deflection.

Solution: Substituting in Fr. 7-5-1,

$$\Delta = \left(\frac{f}{1000}\right)\left(\frac{L^2}{d}\right) = \left(\frac{24}{1000}\right)\left(\frac{16 \times 16}{12}\right) = 0.51 \text{ in.}$$

Formula 7-3-2 shows that the deflection varies directly as f and indirectly as E and d; that is, the greater the fiber stress, the greater the deflection. Likewise, the greater the modulus of elasticity or the greater the depth of the beam, the smaller the deflection. For this reason, Fr. 7-5-1 gives a slightly smaller deflection than is found by using Fr. 7-3-2, which is predicated on a value of 29,000 ksi for E.

Occasionally we design a simple beam on which the loading does not conform with any of the usual types of loading shown in Fig. 4-27—call it "irregular loading." When such a problem occurs, compute the deflection on the basis that the load is uniformly distributed; then multiply this value by 0.92 as given in the footnote to Table 7-1. This procedure gives the deflection for the irregular loading. Although not exact, the approximate deflection found in this manner is sufficiently accurate for average conditions.

7-6 Deflections Found by Coefficients

A method of computing deflections preferred by some designers consists in using the coefficients found in Table 7-2. This table is applicable for simple beams with uniformly distributed loads and for symmetrical sections such as I-beams and wide flange sections.

To use this table, find the coefficient corresponding to the beam span and f, the bending stress. This coefficient divided by the depth of the beam in inches gives the deflection in inches:

$$\text{Deflection} = \frac{\text{coefficient in table}}{\text{depth of beam}}$$

TABLE 7-2. Deflection Coefficients for Uniformly Distributed Loads

Span, feet	Bending stress pounds per square inch		Span, feet	Bending stress pounds per square inch		Span, feet	Bending stress pounds per square inch	
	22,000	24,000		22,000	24,000		22,000	24,000
10	2.274	2.481	18	7.365	8.035	26	15.369	16.766
11	2.750	3.000	19	8.206	8.952	27	16.572	18.079
12	3.273	3.571	20	9.094	9.921	28	17.824	19.448
13	3.840	4.190	21	10.025	10.936	29	19.120	20.858
14	4.455	4.860	22	11.002	12.002	30	20.462	22.322
15	5.115	5.580	23	12.027	13.120	31	21.848	23.834
16	5.821	6.350	24	13.094	14.284	32	23.280	25.396
17	6.571	7.169	25	14.209	15.506	33	24.758	27.009

Coefficient for concentrated load at center of span = 0.8 of above values.
Coefficient for triangular loading, apex at center of span = 0.96 of above values.
Coefficient for equal concentrated loads at $\frac{1}{3}$ points = 1.02 of above values.
Coefficient for irregular loading = 0.92 of above values (approximate).

Table 7-2 has been prepared for *uniformly distributed* loads. For other types of loading, the coefficients are multiplied by the various factors given in the footnote to the table.

Example 1. A W 14 × 34 supports a distributed load of 39,000 lb on a span of 20 ft. The bending stress is 24,000 psi. Determine the deflection by the use of Table 7-2.
Solution: From Table 1-1, we find the depth of this beam to be 14 in.; and in Table 7-2, we see that the coefficient is 9.921. Therefore,

$$\text{deflection} = \frac{9.921}{14} = 0.71 \text{ in.}$$

It may happen that the extreme fiber stress is neither 22,000 psi nor 24,000 psi, as shown in the table. This presents no difficulty, for we know that the deflection is directly proportional to the extreme fiber stress. Therefore, take the coefficient corresponding to one of the two stresses given in the table and reduce or increase it by the ratio of the actual stress to the tabular stress.

Example 2.   A simple beam has a span of 16 ft and a distributed load of 40,000 lb. The section used to support this load is a W 16 × 36. Compute the deflection.

Solution: From Table 1-1, we find that $d = 15.85$ in. and $S = 56.5$ in.^{3} for this section. The maximum bending moment ($M = WL/8$) is computed to be 960,000 in-lb. Then

$$f = \frac{M}{S} = \frac{960,000}{56.5} = 17,000 \text{ psi}$$

is the actual extreme fiber stress.

The coefficient in Table 7-2 for a span of 16 ft 0 in. and an extreme fiber stress of 22,000 psi is 5.821. Then

$$\text{deflection} = \frac{17,050}{22,000} \times \frac{5.821}{15.85} = 0.28 \text{ in.}$$

8

Beam
Design
Procedures

||||||||||||||||||||||||||||||||||||||||||||||||

8-1 General

The complete design of a beam includes consideration of bending
strength, shear resistance, deflection, lateral support, and web
crippling. Design for bending was treated in Chapter 6, and methods
for computing deflections were presented in Chapter 7. In this
chapter, the remaining items are considered, and design procedures
are established.

The current AISC Specification includes special factors that
influence design for bending. The existence of higher strength steels
and attendant higher allowable stresses has led to the classification
of structural shapes as *compact* or *noncompact* sections. Consequently, the shape employed, the laterally unsupported length of
span, and the grade of steel must be known in order to determine
the beam size. Unless otherwise stated, the examples and problems
presented in this book are based on the assumption that A36 steel is
used. The design procedures and specification formulas, however,

are applicable to any grade of steel by selecting the appropriate value of the yield stress, F_y.

8-2 Compact and Noncompact Sections

A compact section is one in which (a) the *width-thickness ratio* of the projecting compression flange (half-flange) does not exceed $52.2/\sqrt{F_y}$; and (b) the *depth-thickness ratio* of the web does not exceed $412/\sqrt{F_y}$. Since the second of these criteria is not critical for W, S, and M shapes in A36 steel, the depth-thickness ratio of the web (d/t_w) will not be considered further in this text. The limiting value of the width-thickness ratio of the compression half-flange ($b_f/2t_f$) is computed as follows for A36 steel:

$$\frac{b_f}{2t_f} = \frac{52.2}{\sqrt{F_y}} = \frac{52.2}{\sqrt{36}} = 8.70$$

To show how the shapes are identified, refer to Fig. 8-1. Two beam sections are shown with their flange and web dimensions as given in Table 1-1. By these dimensions, the width-thickness ratio of the compression half-flange of the W 8 × 28 is

$$\frac{b_f}{2t_f} = \frac{6.54}{2 \times 0.463} = 7.06$$

Because this value is less than the limiting value of 8.70, the W 8 × 28 is a *compact section* for A36 steel.

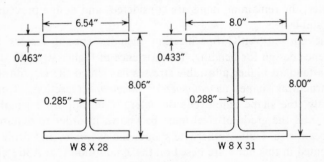

FIGURE 8-1

Now, making the same test for the W 8 × 31,

$$\frac{b_f}{2t_f} = \frac{8.0}{2 \times 0.433} = 9.24 > 8.70$$

Since the value of 9.24 exceeds the limiting value of 8.70, the W 8 × 31 is a *noncompact section*. This classification of beams is a controlling factor in determining the allowable bending stress. For A36 steel, $F_b = 0.66F_y$ or 24 ksi for compact sections. For noncompact shapes, F_b may be taken as $0.60F_y$ or (for A36 steel) 22 ksi.[1]

In order to determine whether a section is compact or noncompact, it is not necessary to make the arithmetical test demonstrated above. Reference to Tables 1-1 and 1-2 will show that values for $b_f/2t_f$ are given under Compact Section Criteria for each of the structural shapes listed. When working with A36 steel, a section will be identified as compact if the tabular value of this ratio *does not exceed* 8.70. This will be the case for most W and for all S sections in A36 steel.

8-3 Lateral Support of Beams

A beam may fail by sidewise buckling of the top (compression) flange when lateral deflection is not prevented. The tendency to buckle increases as the compressive bending stress in the flange increases and also as the unbraced length of span increases. Consequently, the full value of the allowable extreme fiber stress, $F_b = 0.66F_y$, can be used only when the compression flange is adequately braced. To achieve this condition, the AISC Specification requires that the compression flange shall be supported laterally at intervals not to exceed

$$\frac{76.0b_f}{12\sqrt{F_y}} \quad \text{nor} \quad \frac{20,000}{12(d/A_f)F_y}$$

---

[1] However, the current AISC Specification provides a formula for graded reduction of the allowable bending stress from 24 ksi to 22 ksi. The AISC formula (1.5-5) is given in Section 1.5.1.4.2 of the 1969 Specification and may be used when $b_f/2t_f$ exceeds $52.2/\sqrt{F_y}$ but is less than $95.0/\sqrt{F_y}$. Tables to facilitate use of the formula are given in Appendix A of the Specification.

whichever is smaller. The smaller value determined from these expressions is denoted L_c and is expressed in feet. For A36 steel, $F_b = 24$ ksi when the unbraced length does not exceed L_c. Values of L_c are not given in Tables 1-1 and 1-2, "Properties for designing," since they depend upon the grade of steel employed, represented by F_y, as well as the dimensions of the section. However, L_c values are given in connection with safe load tables (discussed in Art. 8-7) and are recorded in the third column of Table 8-2 for several of the W and S Shapes that are listed in Tables 1-1 and 1-2.

The symbol L_u, tabulated adjacent to L_c in Table 8-2, is the maximum unbraced length of the compression flange beyond which the allowable bending stress is less that $0.60F_y$ or, for A36 steel, 22 ksi. For most of the shapes listed in Table 8-2, L_u is determined from the expression

$$L_u = \frac{556}{12(d/A_f)}$$

Values of d/A_f for W and S shapes are given in Tables 1-1 and 1-2.

It is not always a simple matter to decide when a beam is laterally supported. In cases such as that shown in Fig. 8-2a, lateral support is supplied to beams by the floor construction; it is evident from the figure that lateral deflection of the top flange is prevented by the concrete slab. On the other hand, the type of floor construction shown in Fig. 8-2b, where wood joists simply rest on the top flange of an I-beam, offers no resistance to sidewise buckling. Floor systems of the types indicated in Fig. 8-2c, d, e, and f are considered to furnish adequate lateral support of the top flange. However, metal or precast floor systems held in place by clips are not generally considered to have sufficiently rigid connections to the flange to provide adequate lateral bracing. If a beam acts as a girder and supports other beams with connections like those shown in Fig. 11-8, lateral bracing is provided at the connections, and the laterally unsupported length of the top flange to be checked against L_c or L_u becomes the distance between framing points. The design of beams in which the laterally unsupported length exceeds L_u is discussed in Art. 8-10.

Problem 8-3-A. Determine whether a W 12 × 40 is a compact section for A36 Steel.

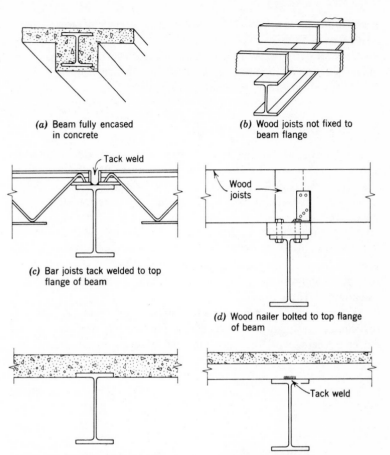

(a) Beam fully encased in concrete

(b) Wood joists not fixed to beam flange

(c) Bar joists tack welded to top flange of beam

Tack weld

(d) Wood nailer bolted to top flange of beam

Wood joists

(e) Beam flange encased in concrete

(f) Metal deck tack welded to top flange of beam

Tack weld

FIGURE 8-2

*Problem 8-3-B.** Is a W 12 × 40 compact for a grade of steel having a yield stress of 50 ksi?

Problem 8-3-C.   For a W 40 × 38 made of A 36 steel, what is the maximum unbraced length of compression flange for an allowable bending stress of (a) 24 ksi and (b) 22 ksi?

8-4 Checking the Shear Stress

It is seldom that the shearing stresses in a beam are a factor in determining its size. It is customary to determine first the size of the beam to resist bending stresses. Having done this, the beam is then investigated for shear, which means that we compute the actual maximum unit shearing stress to see that it does not exceed the allowable stress given in the governing building code. The AISC Specification gives F_v, the allowable shearing stress, as $0.40F_y$ on the gross section of the web of rolled beams. For A36 steel, $F_v = 0.40 \times 36$ or 14.4 ksi, and this value is rounded off at $F_v = 14.5$ ksi, as shown in Table 3-2.

As noted in Art. 2-5, the shearing stresses in beams are not distributed uniformly over the cross section but are zero at the extreme fibers, with the maximum value occurring at the neutral surface. Consequently, the material in the flanges of wide flange sections, I-beams, and channels has little influence on shearing resistance, and the working formula for determining shearing stress is taken as

$$f_v = \frac{V}{A_w}$$

where f_v = the unit shearing stress,
   V = the maximum vertical shear,
and A_w = the gross area of the web (actual depth of section times the web thickness, or $d \times t_w$).

The shearing stresses in beams are seldom excessive. If, however, the beam has a relatively short span with a large load placed near one of the supports, the bending moment is relatively small and the shearing stress becomes comparatively high. This situation is demonstrated in the following example.

Example. A simple beam of A36 steel is 6 ft long and has a concentrated load of 36 kips applied 1 ft from the left end. It is found that a W 8 × 24 is large enough to support this load with respect to bending stresses (required $S = 15$ in.^{3}). Investigate the beam for shear, neglecting the weight of the beam.

Solution: By the methods explained in Art. 4-6, the two reactions are computed; we find that the left reaction $R_1 = 30$ kips and $R_2 = 6$ kips. The maximum vertical shear is, therefore, 30 kips.

We find in Table 1-1 that $d = 7.93$ in. and $t_w = 0.245$ in. for the W 8 × 24, making $A_w = 7.93 × 0.245$ or 1.94 sq in. Then

$$f_v = \frac{V}{A_w} = \frac{30}{1.94} = 15.4 \text{ ksi}$$

the actual unit shearing stress. This, however, exceeds the allowable value of 14.5 ksi, and the W 8 × 24 is not acceptable.

Recalling from Art. 1-3 that standard I-beams have relatively thicker webs than wide flange sections, it is found from Table 1-2 that an S 8 × 23 has a depth of 8 in. and a web thickness of 0.441 in., giving an A_w of 3.5 sq in. Then

$$f_v = \frac{V}{A_w} = \frac{30}{3.53} = 8.53 \text{ ksi}$$

which is less than the allowable 14.5 ksi. It should be observed that S for the S 8 × 23 is 16.2 in.^{3}; for the W 8 × 24, it is 20.8 in.^{3} Since these values are greater than the 15 in.^{3} required, both sections are large enough to resist the bending stresses; but only the S 8 × 23 will provide adequate shearing resistance.

Problem 8-4-A. * Compute and compare the maximum permissible web shears for the following sections in A36 steel: (a) S 12 × 40.8, (b) W 12 × 40, (c) W 10 × 21, (d) C 10 × 20.

8-5 General Design Procedure

The following general procedure gives the usual steps required for the complete design of beams. As one becomes proficient in structural work, the procedure may be abbreviated since the designer will develop an insight with respect to situations where shear, deflection, or lateral support tests may be omitted.

Step 1: Compute the loads the beam will be required to support, and make a sketch showing the loads and their locations. Note the distance between points of lateral support.

Step 2: Compute the reactions.

Step 3: Compute the maximum bending moment.

Step 4: Determine the required section modulus by use of the beam formula $S = M/F_b$, where M is the bending moment in *inch-pounds* or *kip-inches* and F_b is the allowable bending stress in psi or ksi, respectively.

Step 5: Refer to the tables of structural shapes that give properties for designing, and select a beam with a section modulus equal to or greater than that required by Step 4. The beam selected will be adequate for bending stresses subject to checking of section classification (compact or noncompact) and L_c or L_u controls.

Step 6: Check the beam for shear by comparing the computed shearing stress $f_v = V/A_w$ with the allowable F_v.

Step 7: Compute the maximum deflection caused by the loading and compare it with the allowable Δ. If the computed Δ exceeds the allowable, the applicable deflection formula from Fig. 4-27 may be solved for I required to limit Δ to the allowable value.[2] A beam section with an adequate moment of inertia may then be selected from the tables giving properties for designing.

Commentary on Procedure. In computing the loads in Step 1, consideration must be given to the weight of the beam; its weight constitutes a uniformly distributed load. Some designers ignore the beam weight when computing the loads but select a beam with a section modulus somewhat larger than that computed in Step 4. This additional allowance is to provide for the weight of the beam.

A diagram of the beam with its loads and their positions is very important. It shows graphically all the conditions and is an aid to computing the maximum bending moment. When making this diagram, always show the magnitude of the reactions; they are needed subsequently to design bearing plates or connections to girders and columns.

---

[2] For example, the deflection formula for Case 2 of Fig. 4-27 may be transposed to read $I = 5Wl^3/384\,E\Delta$.

The majority of beams encountered in building construction have symmetrical loading and correspond to the typical cases in Fig. 4-27. For such beams, the value of the maximum bending moment, as required in Step 3, is readily found by use of the formulas $M = WL/8$, $M = PL/4$, etc., as given for the individual cases in Fig. 4-27. If, however, the loading is not symmetrical, the bending moment must be computed as explained in Chapter 4. For such cases, we cannot determine the magnitude of the maximum bending moment until we know its position. This, however, presents no difficulty. We merely construct a shear diagram, and the section at which the shear passes from a plus to a minus quantity, "where the shear passes through zero," is the section at which the bending moment is maximum.

Particular attention is called to the necessity of expressing the bending moment in inch-pounds or kip-inches, not foot-pounds or kip-feet, when executing Step 4 of the procedure. This is an error frequently made. If the bending moment is computed in foot-pounds or kip-feet in Step 3, it is readily converted for use in the beam formula by multiplying by 12.

The allowable extreme fiber stress, F_b, to be used in the flexure formula is dependent on several items. First, the type of steel is determined. A summary of the allotted yield strengths and certain of these stresses are given in Table 3-2. Next, the distance between points of lateral support of the beam is established.

For A36 steel, F_b equals 24,000 psi if the beams are compact and are supported laterally at intervals not greater than L_c. For this condition, we sometimes use the expression "adequately braced." If the distance between lateral supports is greater than L_c but not greater than L_u, the allowable bending stress F_b equals 22,000 psi. See Art. 8-3. When the distances between points of lateral support exceed L_u, beams may be designed by the procedure explained in Art. 8-10.

Step 5 is the selection of the proper beam section. If no allowance has been made in Step 1 for the weight of the beam, a beam with a section modulus somewhat larger than that computed in Step 4 is selected. Sometimes architectural considerations may limit the depth of the beam. When, however, there are no limiting conditions, the lightest section having the required section modulus is the most economical section to use. Bear in mind that the beams of greater depth result in smaller deflections. In selecting the beam, Table 8-2 will be found very useful; it will be noted in the second column of the table that the beams are listed in the order of increasing section moduli.[3]

The design procedure did not include a step covering investigation for web crippling. This is a type of failure where an excessive concentrated load on a beam or an excessive end reaction may cause crippling or buckling of the beam web. For average conditions, web stresses in rolled sections seldom exceed the allowable; but when they do, beam webs must be stiffened. The computations required to check this situation are explained in Art. 8-15.

8-6 Examples Illustrating Beam Design

The following examples illustrate the design of various types of beams and loading patterns. To avoid needless repetition, the design procedure outlined in Art. 8-5 is abbreviated in certain instances by omitting the investigation of shear and deflection.

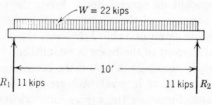

FIGURE 8-3

Example 1.   A simple beam has a span of 10 ft and carries a uniformly distributed load of 22 kips, including its own weight. The deflection is limited to $\frac{1}{360}$ of the span for the total load, and the beam is laterally supported for its entire length. Design the beam in A36 steel using the full procedure given in Art. 8-5.

Solution: (1) A sketch of the beam and loading is made as shown

[3] This arrangement and the listing of corresponding values of L_c and L_u is presented in the AISC Manual in a much more extensive table called Allowable Stress Design Selection Table for Shapes Used as Beams.

in Fig. 8-3. No weight allowance is made for the beam because, by data, its weight is included in the 22 kips.

(2) Reactions: since the beam is symmetrically loaded, $R_1 = R_2 = 22/2 = 11$ kips. This value is recorded on the beam diagram.

(3) The maximum bending moment is given by the formula for Case 2, Fig. 4-27.

$$M = \frac{WL}{8} = \frac{22 \times 10}{8} = 27.5 \text{ kip-ft}$$

(4) Section modulus: since full lateral support is provided, the allowable bending stress is $F_b = 24$ ksi if a compact section is used. Then

$$S = \frac{M}{F_b} = \frac{27.5 \times 12}{24} = 13.75 \text{ in.}^3$$

(5) Referring to Table 1-1 or column two of Table 8-2, we see that a W 8 × 17 has a section modulus of 14.1 in.³ and consequently is accepted as a trial beam.

The compact section criteria in Table 1-1 shows that $b_f/2t_w = 8.52$ for this section. Since this value is less than 8.70, the limiting value for A36 steel (Art. 8-2), the section is compact and the value of $F_b = 24$ ksi used in Step 4 is confirmed.

The value of 5.5 ft for L_c given in Table 8-2 is not significant here since the W 8 × 17 has, by data, full lateral support over the 10-ft span.

(6) Checking the shear stress requires reference to Table 1-1, which shows that $d = 8$ in. and $t_w = 0.23$ in. for the W 8 × 17. Then, since $V = 11$ kips,

$$f_v = \frac{V}{A_w} = \frac{11}{8 \times 0.23} = 5.97 \text{ ksi}$$

This value is less than the allowable 14.5 ksi, so the W 8 × 17 is acceptable for shear.

(7) The allowable deflection is stated in the data as ⅟₃₆₀ of the span and is $(10 \times 12)/360$ or 0.33 in. Since the allowable bending stress for this beam is 24 ksi, Formula 7-3-3, given in Art. 7-3, may be used to find the actual deflection. Then

$$\Delta = \frac{0.02483L^2}{d} = \frac{0.02483 \times 10 \times 10}{8} = 0.31 \text{ in.}$$

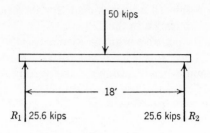

FIGURE 8-4

This value is less than the allowable 0.33 in., so the W 8 × 17 is acceptable for this loading.

Example 2.   A simply supported girder has a span of 18 ft, with a concentrated load of 50 kips at the center of the span. The girder is braced laterally at midspan by the beams framing into it, whose reactions constitute the load. Design the girder for bending stresses using A36 steel.

Solution: (1) A diagram of the beam and loading is made as shown in Fig. 8-4. The distance between points of lateral support is 9 ft. We shall consider the matter of beam weight later.

(2) The reactions due to the superimposed load are each equal to $50/2 = 25$ kips. Each of these will be increased by half the girder's weight when it has been determined.

(3) The maximum bending moment due to the superimposed load (Case 1, Fig. 4-27) is

$$M = \frac{PL}{4} = \frac{50 \times 18}{4} = 225 \text{ kip-ft or } 2700 \text{ kip-in.}$$

(4) Section modulus: since the lateral unbraced length is 9 ft, it is possible that L_c for the beam selected will be less than this value. Therefore, assume as a trial that $F_b = 22$ ksi rather than 24 ksi. Then

$$S = \frac{M}{F_b} = \frac{2700}{22} = 123 \text{ in.}^3$$

(5) The lightest-weight section shown in Table 8-2 with a section modulus equal to or greater than 123 in.^{3} is a W 21 × 62; its S is

127 in.^{3} The margin of $127 - 123 = 4$ in.^{3} may be sufficient to cover the section modulus required by the beam weight.

Making this test, the bending moment due to beam weight is

$$M_{wt} = \frac{wL^2}{8} = \frac{62 \times 18 \times 18 \times 12}{8} = 30.2 \text{ kip-in.}$$

Adding this to the moment developed by the superimposed load, the revised section modulus required is

$$S = \frac{2700 + 30}{22} = \frac{2730}{22} = 124 \text{ in.}^3 < 127 \text{ in.}^3$$

so the W 21 × 62 is acceptable, subject to checking L_c and L_u controls.

Turning to Table 8-2, we note that L_c for the W 21 × 62 is 8.7 ft, and L_u is 11.2 ft. Since the unbraced length of 9 ft falls between these values, the bending stress of 22 ksi assumed in Step 4 is accepted. (It should be noted that, had the distance between points of lateral support not exceeded $L_c = 8.7$ ft, we could have used an F_b of 24 ksi.)

The final reactions may now be computed and recorded on the beam diagram made in Step 1. Total beam weight is $62 \times 18 = 1120$ lb or 1.12 kips. Each reaction then equals $25 + 1.2/2 = 25.56$ or 25.6 kips.

Example 3. A simple beam has a span of 16 ft with a uniformly distributed load of 1000 lb per lin ft, including its own weight, over its entire length. In addition, a concentrated load of 8000 lb is applied 4 ft from the right reaction. The beam is laterally supported throughout its length, and the maximum deflection is limited to ½ in. Design the beam using A36 steel.

Solution: (1) The beam diagram is drawn as shown in Fig. 8-5a.

(2) The reactions for this irregular loading are computed as explained in Art. 4-6:

$16R_1 = (1000 \times 16 \times 8) + (8000 \times 4)$ and $R_1 = 10,000$ lb

$16R_2 = (1000 \times 16 \times 8) + (8000 \times 12)$ and $R_2 = 14,000$ lb

(3) In order to compute the maximum bending moment, it is necessary to determine the point along the span where the shear is

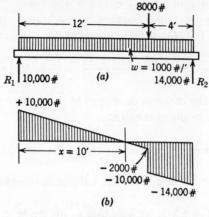

FIGURE 8-5

zero. This may be determined graphically by constructing the shear diagram to scale as in Fig. 8-5b and measuring the distance x from R_1. If the shear diagram is not drawn accurately to scale, x may be found by writing an expression for the shear at this point and equating it to zero.

$$V_x = 10,000 - (1000 \times x) = 0$$

$$x = 10 \text{ ft}$$

Then

$$M_{(x=10)} = (10,000 \times 10) - (1000 \times 10 \times 5)$$
$$= 50,000 \text{ ft-lb} \quad \text{or} \quad 600,000 \text{ in-lb}$$

(4) Since, by data, the beam is laterally supported throughout its length, $F_b = 24,000$ psi; and the required section modulus is

$$S = \frac{M}{F_b} = \frac{600,000}{24,000} = 25 \text{ in.}^3$$

(5) Referring to Table 1-1 or column two of Table 8-2, we see that a W 10 × 25 has an S of 26.5 in.[3]; this section is selected provisionally.

Checking compact section criteria, Table 1-1 shows that $b_f/2t_w = 6.70$. This is less than the limiting value of 8.70, so the section is compact and the value of $F_b = 24,000$ psi used in Step 4 is confirmed.

(6) Checking the shear stress, Table 1-1 gives values of $d = 10.08$ in. and $t_w = 0.252$ in., and

$$f_v = \frac{V}{A_w} = \frac{14,000}{10.08 \times 0.252} = 5500 \text{ psi}$$

which is well within the 14,500 psi allowable.

(7) Since the extreme fiber stress in this beam is approximately 24,000 psi, the deflection may be computed by Formula 7-3-3 for a uniformly distributed load, and the resulting value may be multiplied by the factor 0.92 as given in the footnote to Table 7-1. Then

$$\Delta = 0.92 \times \frac{0.02483L^2}{d} = 0.92 \times \frac{0.02483 \times 16 \times 16}{10.08} = 0.58 \text{ in.}$$

which exceeds the $\frac{1}{2}$ in. allowable. Referring to Table 8-2, the next deeper beam is found to be a W 12 × 27. Table 1-1 shows the depth of this section as 11.96 in. Therefore, the approximate deflection of the W 12 × 27 would be

$$\Delta = 0.58 \times \frac{10.08}{11.96} = 0.49 \text{ in.}$$

which does not exceed the allowable $\frac{1}{2}$ in. Use the W 12 × 27.[4]

Example 4. An overhanging beam with a total length of 24 ft projects 6 ft beyond its right support. There is a uniformly distributed load of 2000 lb per lin ft over its entire length; this includes the weight of the beam. The compression flange will be braced laterally at 6-ft intervals. Design the beam in A36 steel.

Solution: (1) The beam diagram of Fig. 8-6 is first drawn showing the loading and the span dimensions. By data, bracing of the compression flange occurs every 6 ft; but in this overhanging beam, the *bottom* flange will be in compression over the portion of the span subjected to negative bending moment. The deformation curve will be similar to the one shown in Fig. 4-19 (see Art. 4-21).

(2) Computing the reactions,

$$18R_1 = 24 \times 2000 \times 6 \qquad \text{and} \qquad R_1 = 16,000 \text{ lb}$$
$$18R_2 = 24 \times 2000 \times 12 \qquad \text{and} \qquad R_2 = 32,000 \text{ lb}$$

---

[4] The reader should satisfy himself that the W 12 × 27 is adequate for shear and that it is a compact section.

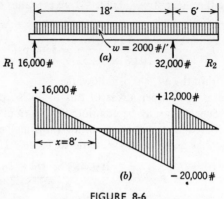

FIGURE 8-6

(3) To find the maximum bending moment, the shear diagram is now drawn (Fig. 8-6b). We note that the shear passes through zero at the right support and also at some point between the two supports. These two points will be the points of maximum negative and maximum positive bending moments. Let x ft·be the distance from R_1 to the point of zero shear. Then

$$16,000 - (2000 \times x) = 0 \qquad \text{and} \qquad x = 8 \text{ ft } 0 \text{ in.}$$

$$M_{(x=8)} = (16,000 \times 8) - (2000 \times 8 \times 4) = 64,000 \text{ ft-lb}$$

which is the maximum positive bending moment.

$$M_{(x=18)} = (16,000 \times 18) - (2000 \times 18 \times 9) = -36,000 \text{ ft-lb}$$

which is the maximum negative bending moment. Since the positive moment is the greater magnitude, it will be used in determining the required section modulus.

(4) To compute the required section modulus, assume an allowable bending stress of 24,000 psi. This will probably be valid in view of the 6-ft intervals between points of lateral bracing; it will be checked later. Then

$$S = \frac{M}{F_b} = \frac{64,000 \times 12}{24,000} = 32 \text{ in.}^3$$

(5) The lightest-weight beam listed in Table 8-2 having a section modulus equal to or greater than 32 in.^{3} is the W 12 × 27 ($S = 34.2$). This table shows that $L_c = 6.9$ ft; and since this is greater

than the bracing interval, the value of $F_b = 24,000$ psi assumed in Step 4 is confirmed.

Note on lateral bracing: Returning to the point raised in Step 1 concerning positive and negative bending moment in this overhanging beam, it is the *bottom* flange of the W 12 × 27 that needs lateral support where negative moment occurs. To locate the position along the span where the bending moment changes from positive to negative requires determination of the inflection point, as discussed in Art. 4-17. Example 2 of Art. 4-17 is similar to this example except that the loads differ by a factor of ten. However, the inflection point in each case occurs as indicated in Fig. 4-25*d*. This means that the bottom flange of the W 12 × 27 is in compression from the end of the overhang to a point 2 ft left of the right reaction. Throughout this 8-ft distance, the lateral bracing must be arranged so that the bottom flange does in fact receive lateral support.

(6) From the shear diagram (Fig. 8-6*b*), we see that the maximum vertical shear is 20,000 lb. Table 1-1 gives dimensions of $d = 11.96$ in. and $t_w = 0.237$ in. for the W 12 × 27. Then

$$f_v = \frac{V}{A_w} = \frac{20,000}{11.96 \times 0.237} = 7050 \text{ psi}$$

Since this value is well within the 14,500 allowable, the W 12 × 27 is accepted.

Example 5. A cantilever beam extends 6 ft beyond the face of its support. The loading consists of a uniformly distributed load (including beam weight allowance) of 500 lb per lin ft and a concentrated load of 6 kips applied at the unsupported end. If the beam is of A36 steel and the compression flange is braced laterally at the unsupported end of the beam, what should be its size?

Solution: (1) The beam diagram is constructed as in Fig. 8-7.

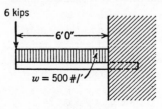

6 kips

6′0″

$w = 500 \#/′$

FIGURE 8-7

(2) The reaction condition is indicated in the beam diagram, and this results in a total vertical shear at the face of the wall of

$$6 \text{ kips} + (0.5 \times 6 = 3 \text{ kips}) = 9 \text{ kips}$$

(3) The maximum bending moment is at the face of the wall and is negative. Then

$$M = (6 \times 6) + (3 \times 3) = 45 \text{ kip-ft}$$

(4) Assuming 24 ksi for F_b, the required section modulus is

$$S = \frac{M}{F_b} = \frac{45 \times 12}{24} = 22.5 \text{ in.}^3$$

(5) From column two of Table 8-2, we find that a W 10 × 25 is the lightest weight beam with adequate section modulus ($S = 26.5$). This table also shows that $L_c = 6.1$ ft, so the value of $F_b = 24$ ksi assumed in Step 4 is confirmed.

(6) As noted in Step 2, the maximum shear for this beam occurs at the face of the support and is 9 kips. From Table 1-1, we find that $d = 10.08$ in. and $t_w = 0.252$ in. for a W 10 × 25. Then

$$f_v = \frac{V}{A_w} = \frac{9}{10.08 \times 0.252} = 3.5 \text{ kips} < 14.5 \text{ kips}$$

The W 10 × 25 is accepted.

Note: In the following problems, assume that A36 steel will be used; and bear in mind that the lightest weight beams are the most economical.

Problem 8-6-A.*   A simple beam has a span of 15 ft. There is a concentrated load of 7.5 kips 3 ft from the left support and a uniformly distributed load of 1.2 kips per lin ft, including beam weight, extending over the full span. The beam is adequately braced laterally. Design the beam.

Problem 8-6-B.*   A beam whose total length is 26 ft overhangs both left and right supports a distance of 4 ft. A uniformly distributed load of 1300 lb per lin ft extends over the full length of the beam. The beam is braced laterally at 6-ft intervals. Design the beam for strength in bending.

Problem 8-6-C.   A beam has a total length of 25 ft and overhangs the left reaction a distance of 5 ft. At the end of the overhang there is a concentrated load of 6 kips, and midway between the two reactions there is a concentrated load of 42 kips. The beam is braced laterally at 5-ft intervals. Design the beam for strength in bending.

Problem 8-6-D. A cantilever beam extends 5 ft beyond the face of its support. At the unsupported end, there is a concentrated load of 8 kips. If the beam is supported laterally throughout its length, what should be its size?

Problem 8-6-E.* A cantilever beam has a length of 4 ft. There is a concentrated load of 10 kips at the unsupported end and a uniformly distributed load (including beam weight) of 1 kip per lin ft extending over the full 4 ft. Lateral support is provided at both ends of the beam. What should be its size?

8-7   Safe Load Tables

The AISC Manual contains tables of total safe uniformly distributed loads, as determined by the allowable bending stress, that simply supported beams can carry on various spans. Such tables are invaluable where there is much designing to be done. Once the total uniformly distributed load has been determined, the proper beam may be selected from the tables without further computations.

Table 8-1, abstracted from the Seventh Edition of the AISC Manual, illustrates the make-up of these tables. The loads tabulated are for A36 steel ($F_y = 36$ ksi) and are based on an allowable bending stress of 24 ksi for compact sections adequately braced. By adequately braced, we mean that the beams are laterally supported at intervals not exceeding L_c. The loads in the table include the weight of the beam; therefore, to find the allowable superimposed load, the weight of the beam must be deducted from the load given in the table.

Loads above the heavy lines in the load columns are limited by maximum allowable web shear ($F_v = 14.5$ ksi). The deflections given in the column at the far right occur when the beams are supporting the full tabulated uniform load.

In addition to listing the section modulus for ready reference, the following data are given at the bottom of the table: V = maximum web shear; R = maximum end reaction for $3\frac{1}{2}$ in. bearing; R_i = increase in R for each additional inch of bearing; N_e = length of end bearing to develop V.

W 10

TABLE 8-1. Typical Beam Safe Load Table**
W SHAPES
Allowable uniform loads in kips
for beams laterally supported

$F_y = 36$ ksi

Designation	W10				W10			
Weight per ft.	66	60	54	*49	45	39	*33	Deflection inches
Flange width	10⅛	10⅛	10	10	8	8	8	
L_c	10.7	10.6	10.6	10.6	8.5	8.4	8.4	
L_u	33.8	31.1	28.4	25.9	22.7	19.6	16.4	

Span in feet	66	60	54	*49	45	39	*33	Deflection inches
6							83	0.09
7					103	92	79	0.12
8	138	123	108	99	98	84	69	0.16
9	131	119	107	96	87	75	61	0.20
10	118	107	97	86	79	68	55	0.25
11	107	98	88	78	71	61	50	0.30
12	98	89	81	72	65	56	46	0.36
13	91	83	74	66	60	52	42	0.42
14	84	77	69	62	56	48	39	0.49
15	79	72	64	57	52	45	37	0.56
16	74	67	60	54	49	42	34	0.64
17	69	63	57	51	46	40	32	0.72
18	66	60	54	48	44	38	31	0.80
19	62	57	51	45	41	36	29	0.90
20	59	54	48	43	39	34	28	0.99
21	56	51	46	41	37	32	26	1.09
22	54	49	44	39	36	31	25	1.20
23	51	47	42	37				
24	49	45	40	36				
25	47	43	39	34				
26	45	41	37					
27	44	40	36					
28	42	38	35					
29	41	37						
30	39	36		(23.7)			(23.6)	

Properties and reaction values

	66	60	54	*49	45	39	*33
S in.³	73.7	67.1	60.4	54.6	49.1	42.2	35.0
V kips	69	62	54	49	51	46	41
R kips	59	53	47	42	44	40	35
R_i kips	12.3	11.2	9.9	9.2	9.5	8.6	7.9
N_e in.	4.3	4.3	4.2	4.2	4.2	4.2	4.2

$F_y = 36$ ksi

Load above heavy line is limited by maximum allowable web shear.
* Tabulated loads for this shape are computed with the allowable stress (ksi) shown in parentheses at the bottom of the allowable load column.
** Taken from a more complete set of tables in the 7th Edition of the *Manual of Steel Construction*. Courtesy American Institute of Steel Construction.

Since it is not possible to reproduce extensive safe load tables within the scope of this book, Table 8-2 has been compiled from data in the AISC Manual to provide a partial working safe load table for several selected beam sections. This table, from which data on L_c and L_u were cited earlier in the chapter, is for use with compact beams that are laterally braced at intervals not greater than L_c. For beams supported laterally at intervals greater than L_c but not greater than L_u, the allowable bending stress is 22 ksi. If this lower stress is to be used, the loads shown in the table must be multiplied by $^{22}\!/_{24}$, a reduction of $8\frac{1}{2}\%$.

In order to use Table 8-2 directly when $F_b = 22$ ksi, we multiply the uniform *load to be carried* by 1.09. (This is equivalent to reducing all the loads given in the table by $^{22}\!/_{24}$.) However, care must be exercised when later recording the reactions of the beam since these must reflect the *actual* loading only. Obviously, this manipulation applies to the loads as controlled by bending requirements only; the designer must determine separately whether the shear and deflection specifications are satisfied by the beam selected.

As an aid in checking deflection, heavy vertical lines have been incorporated in Table 8-2. Tabulated loads to the right of these lines will produce deflections exceeding $\frac{1}{360}$ of the span.

Example 1. Using the data given for Example 1 of Art. 8-6, select the beam from the safe load table (Table 8-2).
Solution: Since A36 steel is to be used and full lateral support is provided, $F_b = 24$ ksi, and Table 8-2 can be used directly. Entering the 10-ft span column, the first load equal to or greater than 22 kips is 22.6 kips listed for the W 8 × 17. This is the same beam selected by the section modulus method in Steps 4 and 5 of the solution of Example 1, Art. 8-6. Since the tabular load appears to the left of the heavy vertical line, we know that the deflection does not exceed $\frac{1}{360}$ of the span.

Example 2. A beam has a span of 14 ft with a uniformly distributed load of 35.7 kips extending over its full length. If the allowable bending stress is 22 ksi, what should be the size of the beam?
Solution: In order to use Table 8-2 for this problem, we multiply the load to be carried by 1.09. Then 35.7 × 1.09 = 39 kips. Entering the 14-ft span column of the table, we find that a load of 39.1 kips

TABLE 8-2. Allowable Uniform Loads in kips for Selected W and S Shapes Used as Beams Laterally Supported, Based on $F_y = 36$ ksi*

Shape	S	L_c	L_u	Span in feet						
	in.³	ft	ft	8	9	10	11	12	13	14
W 8 × 17	14.1	5.5	9.4	28.2	25.1	22.6	20.5	18.8	17.4	16.1
S 8 × 18.4	14.4	4.2	9.9	28.8	25.6	23.0	20.9	19.2	17.7	16.5
S 8 × 23	16.2	4.4	10.3	32.4	28.8	25.9	23.6	21.6	19.9	18.5
W 8 × 20	17.0	5.6	11.3	34.0	30.2	27.2	24.7	22.7	20.9	19.4
W 8 × 24	20.8	6.9	15.1	41.6	37.0	33.3	30.3	27.7	25.6	23.8
W 10 × 21	21.5	6.1	9.1	43.0	38.2	34.4	31.3	28.7	26.5	24.6
W 8 × 28	24.3	6.9	17.4	48.6	43.2	38.9	35.3	32.4	29.9	27.8
S 10 × 25.4	24.7	4.9	10.6	49.4	43.9	39.5	35.9	32.9	30.4	28.2
W 10 × 25	26.5	6.1	11.4	53.0	47.1	42.4	38.5	35.3	32.6	30.3
S 10 × 35	29.4	5.2	11.2	58.8	52.3	47.0	42.8	39.2	36.2	33.6
W 10 × 29	30.8	6.1	13.2	61.6	54.8	49.3	44.8	41.4	37.9	35.2
W 12 × 27	34.2	6.9	10.1	68.4	60.8	54.7	49.7	45.6	42.1	39.1
S 12 × 31.8	36.4	5.3	10.5	73	65	58	53	49	45	42
S 12 × 35	38.2	5.4	10.7	76	68	61	56	51	47	44
W 12 × 31	39.5	6.9	11.6	79.0	70.2	63.2	57.5	52.7	48.6	45.1
W 14 × 30	41.9	7.1	8.6	83	74	66	60	55	51	47
W 10 × 39	42.2	8.4	19.6	84	75	68	61	56	52	48
S 12 × 40.8	45.4	5.5	13.4	91	81	73	66	61	56	52
W 12 × 36	46.0	6.9	13.4	92.0	81.8	73.6	66.9	61.3	56.6	52.6
W 14 × 34	48.6	7.1	10.1	97	86	78	71	65	60	56
S 12 × 50	50.8	5.8	13.9	102	90	81	74	68	63	58
W 12 × 40	51.9	8.4	16.0	102	92	83	75	69	64	59
W 14 × 38	54.7	7.2	11.4	109	97	88	80	73	67	63
W 16 × 36	56.5	7.4	8.7	113	100	90	82	75	70	65
W 12 × 45	58.2	8.5	17.8	116	103	93	85	78	72	67
S 15 × 42.9	59.6	5.8	10.6	119	106	95	87	79	73	68
W 14 × 43	62.7	8.4	14.3	122	111	100	91	84	77	72

will be safely carried by a W 12 × 27. Since the solid vertical line is well to the right of the 14-ft span column, the deflection is well within the $\frac{1}{360}$ limit.

Example 3. A simple beam, adequately braced, has a span of 20 ft and supports a total uniformly distributed load of 33 kips. This full load is composed of a dead load, due to the weight of the construction, of 13 kips and a live load of 20 kips. The beam used to support this load is a W 14 × 30. If the deflection due to the *live*

15	16	17	18	19	20	21	22	23	24	25	26
15.0	14.1	13.3									
15.4	14.4	13.6									
17.3	16.2	15.2									
18.1	17.0	16.0									
22.2	20.8	19.6									
22.9	21.5	20.2	19.1	18.1	17.2	16.4	15.6				
25.9	24.3	22.9									
26.3	24.7	23.2	22.0	20.8	19.8	18.8					
28.3	26.5	24.9	23.6	22.3	21.2	20.2	19.3				
31.4	29.4	27.7	26.1	24.8	23.5	22.4					
32.9	30.8	29.0	27.4	25.9	24.6	23.5	22.4				
36.5	34.2	32.2	30.4	28.8	27.4	26.1	24.9	23.8	22.8	21.9	
39	36	34	32	31	29	28	26	25	24	23	
41	38	36	34	32	31	29	28	27	25	24	
42.1	39.5	37.2	35.1	33.3	31.6	30.1	28.7	27.5	26.3	25.3	
44	41	39	37	35	33	32	30	29	28	27	25
45	42	40	38	36	34	32	31				
48	45	43	40	38	36	35	33	32	30	29	
49.1	46.0	43.3	40.9	37.8	36.8	35.0	33.5	32.0	30.7	29.4	
52	49	46	43	41	39	37	35	34	32	31	30
54	51	48	45	43	41	39	37	35	34	33	
55	52	49	46	44	42	40	38	36	35	33	
58	55	51	49	46	44	42	40	38	36	35	34
60	57	53	50	48	45	43	41	39	38	36	35
62	58	55	52	49	47	44	42	40	39	37	
64	60	56	53	50	48	45	43	41	40	38	37
67	63	59	56	53	50	48	46	44	42	40	39

Loads to right of heavy vertical lines produce deflections exceeding $\frac{1}{360}$ of the span.
* Compiled from data in the 7th Edition of the *Manual of Steel Construction.* Courtesy American Institute of Steel Construction.

load is limited to $\frac{1}{360}$ of the span, will the deflection of this beam be excessive?

Solution: Referring to Table 8-2, we find that a W 14 × 30 will support a uniform load of 33 kips on a span of 20 ft. However, it is observed from the position of the heavy vertical line that the deflection under the *full* load will exceed $\frac{1}{360}$ of the span. It remains to

TABLE 8-2. Allowable Uniform Loads in kips for Selected W and S Shapes Used as Beams Laterally Supported, Based on $F_y = 36$ ksi (Continued)

Shape	S	L_c	L_u	Span in feet						
	in.³	ft	ft	15	16	17	18	19	20	21
W 16 × 40	64.6	7.4	10.2	69	65	61	57	54	52	49
W 12 × 50	64.7	8.5	19.7	69	65	▌ 61	58	54	52	49
W 14 × 48	70.2	8.5	16.0	75	70	66	62	▌ 59	56	53
W 16 × 45	72.5	7.4	11.4	77	73	68	64	61	58	55
W 14 × 53	77.8	8.5	17.6	83	78	73	69	▌ 66	62	59
W 18 × 50	89.1	7.9	11.0	95	89	84	79	75	71	68
S 18 × 54.7	89.4	6.3	10.7	95	89	84	79	75	72	68
W 16 × 58	94.4	8.9	15.9	101	94	89	84	79	76	72
W 18 × 55	98.4	8.0	12.1	105	98	93	87	83	79	75
S 18 × 70	103	6.6	11.1	110	103	97	92	87	82	78
W 16 × 64	104	9.0	17.6	111	104	98	92	88	83	79
W 21 × 55	110	8.7	9.5	117	110	104	98	93	88	84
W 18 × 64	118	9.2	15.5	126	118	111	105	99	94	90
W 21 × 62	127	8.7	11.2	135	127	120	113	107	102	96
W 18 × 70	129	9.2	16.9	138	129	121	115	109	103	98
W 21 × 68	140	8.7	12.4	149	140	132	124	118	112	107
W 24 × 68	153	9.5	10.2	163	153	144	136	129	122	117
W 21 × 82	169	9.5	15.8	180	169	159	150	142	135	129
W 24 × 76	176	9.5	11.9	188	176	166	156	148	141	134
W 18 × 96	185	12.4	24.9	197	185	174	164	156	148	141
W 27 × 84	212	10.5	11.2	226	212	200	188	179	170	162
W 27 × 94	243	10.5	12.8	259	243	229	216	205	194	185
W 24 × 100	250	12.7	17.9	267	250	235	222	211	200	190
W 21 × 112	250	13.7	24.8	267	250	235	222	211	200	190
W 30 × 99	270	10.9	11.6	288	270	254	240	227	216	206
W 24 × 110	276	12.7	19.7	294	276	260	245	232	221	210
W 21 × 127	284	13.8	28.1	303	284	267	252	239	227	216
W 30 × 108	300	11.1	12.4	320	300	282	267	253	240	229
W 24 × 130	332	14.8	24.1	354	332	312	295	280	266	253

determine whether the deflection under the live load alone exceeds this limit, which is $(20 \times 12)/360 = 0.66$ in.

The deflection under live load alone is

$$\Delta = \frac{5}{384} \times \frac{Wl^3}{EI} \quad \text{and} \quad \Delta = \frac{5}{384} \times \frac{20 \times (20 \times 12)^3}{29,000 \times 290} = 0.43 \text{ in.}$$

which is less than the allowable 0.66 in. Thus, we see from Table 8-2

22	23	24	25	26	27	28	29	30	31	32	33
47	45	43	41	40	38	37	36	34	33	32	31
47	45	43	41								
51	49	47	45	43	42	40	39	37			
53	50	48	46	45	43	41	40	39	37	36	35
57	54	52	50	48	46	44	43	41			
65	62	59	**57**	55	53	51	49	48	46	45	43
65	62	60	**57**	55	53	51	49	48	46	45	43
69	66	63	**60**	58	56	54	52	50	49	47	46
72	68	66	**63**	61	58	56	54	52	51	49	48
75	72	69	**66**	63	61	59	57	55	53	52	50
76	72	69	67	64	62	59	57	55	54	52	50
80	76	73	70	68	65	63	**61**	59	57	55	53
86	82	79	**76**	73	70	67	65	63	61	59	57
92	88	85	81	78	75	73	**70**	68	65	64	61
94	90	86	**83**	79	76	74	71	69	67	65	63
102	97	93	90	86	83	80	**77**	75	72	70	68
111	106	102	98	94	91	87	84	82	79	77	**74**
123	118	113	108	104	100	97	**93**	90	87	85	82
128	122	117	113	108	104	101	97	94	91	88	**85**
135	129	123	**118**	114	110	106	102	99	95	93	90
154	147	141	136	130	126	121	117	113	109	106	103
177	169	162	156	150	144	139	134	130	125	122	118
182	174	167	160	154	148	143	138	133	129	125	**121**
182	174	167	160	154	148	143	**138**	133	129	125	121
196	188	180	173	166	160	154	149	144	135	127	120
201	192	184	177	170	164	158	152	147	142	138	**134**
207	198	189	182	175	168	162	**157**	151	147	142	138
218	209	200	192	185	178	171	166	160	155	150	145
241	231	221	212	204	197	190	183	177	171	166	**161**

that the total load (live plus dead) causes a deflection greater than $\frac{1}{360}$ of the span, but the deflection that results from the live load alone is found to be within the prescribed limits.

Example 4. Compute the total safe uniform load for a laterally supported W 10 × 45 of A36 steel on a span of 18 ft.

Solution: (1) The section modulus of a W 10 × 45 is given in

Table 1-1 as 49.1 in.[3] Using form (1) of the beam formula (Art. 6-1), the resisting moment of the section is

$$M = F_b S = 24 \times 49.1 = 1180 \text{ kip-in.}$$

(2) The bending moment under the given loading is expressed by the formula $M = WL/8$ (Fig. 4-27), which may be written $W = 8M/L$.

(3) Substituting the computed value of resisting moment for bending moment in the above equation, the safe load is

$$W = \frac{8M}{L} = \frac{8 \times 1180}{18 \times 12} = 43.7 \text{ kips}$$

It should be noted that L is multiplied by 12 to convert the span length to inches since the resisting moment is stated in kip-in. This value may be checked by referring to Table 8-1.

Note: The following problems are to be solved by the use of safe load tables. In each, assume the beam to be braced laterally. The steel is A36; deflection for the total load is not to exceed $\frac{1}{360}$ of the span; and the allowable bending stress is 24 ksi except as noted. For each problem, select the lightest weight beam that satisfies the requirements.

Problem 8-7-A. With the use of Table 8-2, select the lightest weight beam for a span of 13 ft with a uniformly distributed load of 30 kips.

Problem 8-7-B. A simple beam has a span of 18 ft and a uniformly distributed load of 45 kips. What should be its size?

Problem 8-7-C.* A uniform load of 50 kips is supported by a simple beam which has a span of 13 ft. What beam having the *shallowest depth* will properly support this load?

Problem 8-7-D.* The total uniform load on a beam with a span of 16 ft is 33 kips. By the use of Table 8-2, determine the proper size beam if the allowable bending stress is 22 ksi.

8-8 Safe Load Table for Channels

When American Standard channels (C) are used as beams, the allowable bending stress for A36 steel is 22 ksi, provided the compression flange is braced laterally at intervals not greater than L_u

as determined by

$$L_u = \frac{556}{12(d/A_f)}$$

Values of d/A_f for channels are given in Table 1-3. Safe loads for selected channel sections used as beams are given in Table 8-3, which has been computed in accordance with these provisions.

8-9 Equivalent Tabular Loads

The safe loads shown in Tables 8-1, 8-2, and 8-3 are uniformly distributed loads on simple beams. By the use of coefficients, we can convert other types of loading to equivalent uniform loads and thereby greatly extend the usefulness of the tables.

The maximum bending moments for typical loadings are shown in Fig. 4-27. For a simple beam with a uniformly distributed load, $M = WL/8$. For a simple beam with equal concentrated loads at the third points, $M = PL/3$. These values are shown in Fig. 4-27, Cases 2 and 3, respectively. Equating these values,

$$\frac{WL}{8} = \frac{PL}{3} \quad \text{and} \quad W = 2.67 \times P$$

which shows that, if the value of one of the concentrated loads (in Case 3) were multiplied by the coefficient 2.67, we would have an equivalent distributed load that would produce the same bending moment as the concentrated loads. The coefficients for finding equivalent uniform loads for other beams and loadings are given in Fig. 4-27. Because of their use with safe load tables, equivalent uniform loads are usually called *equivalent tabular loads*, abbreviated *ETL*.

It is important to remember that an *ETL* does not include the weight of the beam, for which an estimated amount should be added. Beams found by this method should be investigated for shear and deflection; it is assumed that they are adequately supported laterally. Also, when recording the beam reactions, these must be determined from the *actual* loading conditions without regard to the *ETL*.

Examples. In the following examples, the proper beam sizes with respect to bending strength have been determined by use of Table 8-2. In order to clarify the procedure, the weight of the beam has

TABLE 8-3.   Allowable Uniform Loads in kips for Selected Channels (C) Used as Beams Laterally Supported, Based on $F_y = 36$ ksi*

Shape	S	L_u	Span in feet							
	in.³	ft	10	11	12	13	14	15	16	17
C 6 × 8.2	4.38	5.1	6.4	5.8	5.4	4.9	4.6	4.3		
C 6 × 10.5	5.06	5.4	7.4	6.7	6.2	5.7	5.3	4.9		
C 6 × 13	5.80	5.7	8.5	7.7	7.1	6.5	6.1	5.7		
C 7 × 9.8	6.08	5.1	8.9	8.1	7.4	6.9	6.4	5.9	5.6	5.2
C 7 × 12.25	6.93	5.3	10.2	9.2	8.5	7.8	7.3	6.8	6.4	6.0
C 7 × 14.75	7.78	5.6	11.4	10.4	9.5	8.8	8.2	7.6	7.1	6.7
C 8 × 11.5	8.14	5.1	11.9	10.9	9.9	9.2	8.5	8.0	7.5	7.0
C 8 × 13.75	9.03	5.3	13.2	12.0	11.0	10.2	9.5	8.8	8.3	7.8
C 9 × 13.4	10.6	5.2	15.5	14.1	13.0	12.0	11.1	10.4	9.7	9.1
C 8 × 18.75	11.0	5.7	16.1	14.7	13.4	12.4	11.5	10.8	10.1	9.5
C 9 × 15	11.3	5.3	16.6	15.1	13.8	12.7	11.8	11.0	10.4	9.7
C 9 × 20	13.5	5.6	19.8	18.0	16.5	15.2	14.1	13.2	12.4	11.6
C 10 × 15.3	13.5	5.3	19.8	18.0	16.5	15.2	14.1	13.2	12.4	11.6
C 10 × 20	15.8	5.5	23.2	21.1	19.3	17.8	16.6	15.4	14.5	13.6
C 10 × 25	18.2	5.8	26.7	24.3	22.2	20.5	19.1	17.8	16.7	15.7
C 10 × 30	20.7	6.1	30.4	27.6	25.3	23.4	21.7	20.2	19.0	17.9
C 12 × 20.7	21.5	5.7	31.5	28.7	26.3	24.3	22.5	21.0	19.7	18.5
C 12 × 25	24.1	5.9	35.3	32.1	29.5	27.2	25.2	23.6	22.1	20.8
C 12 × 30	27.0	6.1	39.6	36.0	33.0	30.5	28.3	26.4	24.8	23.3
C 15 × 33.9	42.0	6.8	61.6	56.0	51.3	47.4	44.0	41.1	38.5	36.2
C 15 × 40	46.5	7.1	68.2	62.0	56.8	52.5	48.7	45.5	42.6	40.1
C 15 × 50	53.8	7.5	78.9	71.7	65.8	60.7	56.4	52.6	49.3	46.4

been neglected; but this additional load must, of course, be given consideration in actual design. In each case, it is intended that the lightest weight beam that will carry the load be selected.

Data.   A simple beam has a span of 18 ft with a concentrated load of 20 kips at the center of the span.

Solution: This loading is an example of Case 1 in Fig. 4-27. We note that the $ETL = 2 \times P$ or $2 \times 20 = 40$ kips. In Table 8-2, we see that an S 12 × 40.8 will support this load but also that a W 14 × 34, which weighs 6.8 lb less per linear foot, will carry a greater load. Accept the W 14 × 34.

Data.   A simple beam has equal concentrated loads of 9 kips each at the third points of a 21-ft span.

18	19	20	21	22	23	24	25	26	27	28	29
5.0											
5.6											
6.3											
6.6	6.3	6.0									
7.4	7.0	6.6									
8.6	8.2	7.8	7.4	7.1	6.8						
9.0	8.5	8.1									
9.2	8.7	8.3	7.9	7.5	7.2						
11.0	10.4	9.9	9.4	9.0	8.6						
11.0	10.4	9.9	9.4	9.0	8.6	8.3	7.9				
12.9	12.2	11.6	11.0	10.5	10.1	9.7	9.3				
14.8	14.0	13.3	12.7	12.1	11.6	11.1	10.7				
16.9	16.0	15.2	14.5	13.8	13.2	12.7	12.1				
17.5	16.6	15.8	15.0	14.3	13.7	13.1	12.6	12.1	11.7	11.3	10.9
19.6	18.6	17.7	16.8	16.1	15.4	14.7	14.1	13.6	13.1	12.6	12.2
22.0	20.8	19.8	18.9	18.0	17.2	16.5	15.8	15.2	14.7	14.1	13.7
34.2	32.4	30.8	29.3	28.0	26.8	25.7	24.6	23.7	22.8	22.0	21.2
37.9	35.9	34.1	32.5	31.0	29.7	28.4	27.3	26.2	25.3	24.4	23.5
43.8	41.5	39.5	37.6	35.9	34.3	32.9	31.6	30.3	29.2	28.2	27.2

Loads to right of heavy vertical lines produce deflections exceeding $\frac{1}{360}$ of the span.
* Compiled from data in the 7th Edition of the *Manual of Steel Construction*. Courtesy American Institute of Steel Construction.

Solution: This is Case 3, Fig. 4-27. $ETL = 2.67 \times P$ or $2.67 \times 9 = 24$ kips. From Table 8-2, select a W 12 × 27.

Data. A cantilever beam has a length of 12 ft and a concentrated load of 5 kips at the unsupported end.
Solution: This loading corresponds to Case 4 of Fig. 4-27. $ETL = 8 \times P$ or $8 \times 5 = 40$ kips. A W 12 × 27 is selected from Table 8-2.

Data. A simple beam has a span of 24 ft and three concentrated loads of 20 kips each located at the quarter points of the span.

Solution: $ETL = 4 \times P$ for Case 5, Fig. 4-27. Then $ETL = 4 \times 20 = 80$ kips. Table 8-2 shows that a W 21 × 62 is acceptable.

Data.   A cantilever beam projects 10 ft beyond its support. It supports a total uniformly distributed load of 6 kips.

Solution: For this type of loading (Case 6, Fig. 4-27), $ETL = 4 \times W$ or $4 \times 6 = 24$ kips. From Table 8-2, select a W 8 × 20.

Problems 8-9-A-B-C-D-E.   Construct problems illustrating typical beam loadings as shown in Cases 1, 3, 4, 5, and 6 of Fig. 4-27. Select the proper beam sizes from the safe load tables using the *ETL* method, and check the accuracy of your selections by computing the required section modulus in each instance.

8-10   Laterally Unsupported Beams

When the flanges of compact beams are supported at intervals not greater than L_c, the allowable bending stress is 24 ksi for A36 steel. If the distance between points of lateral support exceeds L_c but is less than the value of L_u, the allowable bending stress for A36 steel is 22 ksi (except as a higher stress may be permitted in accordance with the AISC formula cited in the footnote to Art. 8-2). For shapes that do not qualify as compact sections, the allowable bending stress for A36 steel is 22 ksi for all laterally unbraced lengths up to L_u.

When the unbraced length of the compression flange exceeds L_u, the allowable bending stress must be reduced in accordance with certain formulas and provisions set forth in the AISC Specification. The design of beams by use of the specified provisions is not a simple matter, and the AISC Manual gives supplementary charts to aid the designer. These beam charts give "Allowable Moments in Beams with Unbraced Lengths Greater than L_u" and provide a workable approach to this otherwise rather cumbersome problem. One of these charts from the 7th Edition of the AISC Manual is reproduced in Fig. 8-8 to demonstrate their nature and method of use.

After determining the maximum bending moment in kip-ft and noting the longest unbraced length of compression flange, these two coordinates are located on the sides of the chart and traced to their

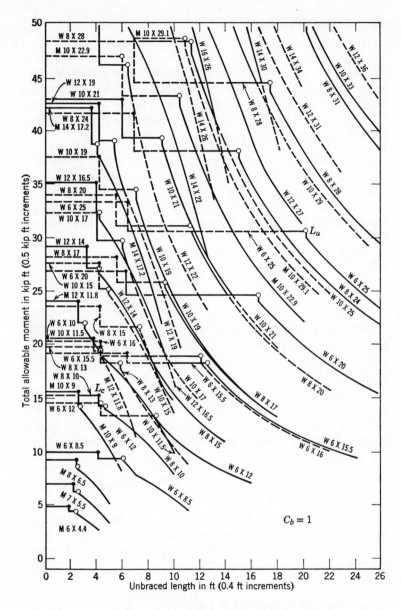

FIGURE 8-8 Allowable moments in beams ($F_y = 36$ ksi). Taken from a more complete set of charts in the 7th Edition of the *Manual of Steel Construction*. Courtesy American Institute of Steel Construction.

intersection. Any beam whose curve lies above and to the right of this intersection satisfies the bending stress requirement. The nearest curve that is a solid line represents the most economical section in terms of beam weight. If there is any question about deflection being critical, it should be investigated; if found excessive, another selection should be made from the chart.

Example. A simple beam carries a total uniform load of 19.6 kips, including its own weight, over a span of 20 ft. It has no lateral support except at the ends of the span. Assuming A36 steel, select from the chart of Fig. 8-8 a beam that will meet bending strength requirements and will not deflect more than 0.75 in. under the full load.

Solution: (1) The maximum bending moment is

$$M = \frac{WL}{8} = \frac{19.6 \times 20}{8} = 49 \text{ kip-ft}$$

(2) Entering the chart on the bottom scale with an unbraced length of 20 ft, proceed vertically to intersection with the horizontal line representing a moment of 49 kip-ft on the scale to the left. The nearest solid-line curve above and to the right of the intersection represents a W 8 × 31.

(3) Observing that this is quite a shallow beam for the 20-ft span length, a deflection check is called for. Since we know the allowable deflection (0.75 in.), it will be convenient to solve the deflection formula for the moment of inertia required to maintain this limit. Referring to Case 2, Fig. 4-27, and making the transformation,

$$I = \frac{5}{384} \times \frac{Wl^3}{E\Delta} = \frac{5 \times 19.6 \times (20 \times 12)^3}{384 \times 29{,}000 \times 0.75} = 162 \text{ in.}^4$$

For the W 8 × 31 selected under Step 2, $I = 110$ in.^{4} (Table 1-1). Since this is less than the 162 in.^{4} required, we return to the chart of Fig. 8-8 and select the W 10 × 33. Table 1-1 shows $I = 171$ in.^{4} for this section, which is adequate. Use the W 10 × 33.

8-11   Long Spans and Light Loads

A problem that occurs very often is the design of a beam whose span is long and whose load is relatively light. The example in Art. 8-10 applies to this situation where the beam is laterally unsupported, but the condition also arises with some of the lighter weight floor construction systems that furnish full lateral support (Fig. 8-2c and d). The following example illustrates this more usual case.

Example. A simple beam adequately braced has a length of 22 ft and supports a total uniform load of 8 kips, including its own weight. Determine the size of the lightest weight steel beam that will be sufficiently strong and whose deflection will not exceed $\frac{1}{360}$ of the span length under the total load; i.e., $(22 \times 12)/360 = 0.74$ in.

Solution: (1) Inspection of Table 8-2 shows that the loads tabulated for a span of 22 ft far exceed 8 kips. Bear in mind that the loads in the table produce a bending stress of 24 ksi. The beam we select will probably have a bending stress considerably smaller.

(2) The maximum bending moment is

$$M = \frac{WL}{8} = \frac{8 \times 22}{8} = 22 \text{ kip-ft or } 264 \text{ kip-in.}$$

and the required section modulus is

$$S = \frac{M}{F_b} = \frac{264}{24} = 11.0 \text{ in.}^3$$

(3) The lightest weight section listed in Table 8-2 which has a section modulus equal to or greater than 11.0 in.[3] is a W 8 × 17, but this beam will have to be tested for deflection (allowable $\Delta = 0.74$ in.). As we have seen before, this test may be accomplished most readily by solving the deflection formula (Case 2, Fig. 4-27) for the I required to maintain the specified limit. Then, if the W 8 × 17 proves inadequate, another beam can be selected directly from the table on the basis of required moment of inertia. Making the transformation and substituting,

$$I = \frac{5}{384} \times \frac{Wl^3}{E\Delta} = \frac{5 \times 8 \times (22 \times 12)^3}{384 \times 29,000 \times 0.74} = 89.3 \text{ in.}^4$$

the required moment of inertia.

(4) From Table 1-1, we find that I for the W 8 × 17 is only 56.6 in.[4], so this section is not adequate. Scanning Table 1-1 further, the lightest weight section with a moment of inertia of 89.3 or larger

is a W 10 × 21 ($I = 107$ in.^{4}). Since the section modulus of this beam is 21.5 in.^{3} and only 11.0 in.^{3} is required, the W 10 × 21 meets the requirements for both strength and deflection.

If the thickness of the floor construction were restricted and a beam 10 in. deep could not be accepted, Table 1-1 shows that a W 8 × 28 has an I of 97.8 in.^{4}, which is also satisfactory. Thus, we see that for long spans and light loads the size of the beam is determined by the moment of inertia rather than by the section modulus; deflection is the controlling factor.

In order to fix this relationship in mind, it will be of interest to determine the actual bending stress developed in both these beams under the full load. For the W 10 × 21, this is

$$f_b = \frac{M}{S} = \frac{264}{21.5} = 12.3 \text{ ksi}$$

and for the W 8 × 28, it is

$$f_b = \frac{M}{S} = \frac{264}{24.3} = 10.9 \text{ ksi}$$

Both of these values are, of course, well within the allowable bending stress of 24 ksi for A36 steel.

Problem 8-11-A.* A simple beam adequately braced laterally has a span of 28 ft and a total live and dead uniformly distributed load of 6 kips. What is the lightest weight steel beam that fulfills the requirements for both strength in bending and deflection, if the latter is limited to $\frac{1}{360}$ of the span under the full load?

8-12 Lintels

When opening such as doorways and windows occur in masonry walls, it is necessary to provide some means of support for the masonry directly above the opening. This may be accomplished by the use of masonry arches or by beams. Beams employed for this purpose are called *lintels*. The load to be supported, the thickness of the wall, and the width of the opening determine the cross section of the lintel that should be used. A number of types of steel lintel are shown in Fig. 8-9. The material to be supported—brickwork, cut-stone, concrete blocks, etc.—and the position of the opening in the

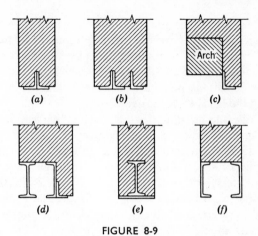

(a)   (b)   (c)

(d)   (e)   (f)

FIGURE 8-9

building govern the makeup of the lintel; steel angle sections are particularly suitable for brickwork.

It is impossible to determine accurately the load to be supported by lintels. The arch-action in the masonry, as may be observed from cracks that sometimes develop, indicate that the loading is approximately triangular in shape. One common practice considers the area of masonry to be supported as a triangle in which the sides make an angle of 60° with the horizontal. This is indicated in Fig. 8-10. Minute accuracy in computing this load is unnecessary. Sometimes floor joists or beams, bearing on the masonry walls, contribute load to a lintel and must be taken into account in the design. The discussion here, however, is directed primarily to situations where the

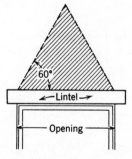

FIGURE 8-10

triangular loading may be assumed—that is, where there is a height of unbroken masonry above the opening about equal to the span length and where there is a substantial pier between the opening under consideration and an adjacent opening. Table 9-1 which lists the weights of various materials of construction will be useful in determining the loads to be carried.

If brickwork is to be supported, the outstanding legs of the angles shown in Fig. 8-9a and b would normally be at least 3½ in. in order to provide an adequate bearing surface on which to lay the bricks. Many designers consider a minimum standard angle for lintels to be an L 3½ × 3 × ⁵⁄₁₆ or an L 3½ × 3 × ⅜ placed with the 3½-in. leg horizontal, regardless of the span. The standard lintel is then investigated to ascertain whether it is sufficiently strong. Such lintels in brick walls should extend beyond the face of the opening a minimum of 4 to 6 in. on each side. Of course, as the length of span increases, the required height of the vertical legs will also increase. For wide openings and heavy loads, lintels such as those shown in Fig. 8-9d, e, and f are employed.

When angles are used as lintels, there is always some uncertainty as to whether the masonry furnishes adequate lateral support to that portion of the vertical leg that is in compression due to bending. Fortunately, the fact that angle lintels are so frequently oversized in order to provide bearing surface for the masonry (thereby resulting in extreme fiber stresses well below the 22 ksi permitted for angles with full lateral support) makes this problem of less consequence than it otherwise would be.

Example 1   A doorway 6 ft 0 in. wide occurs in a 12-in. wall built of common brick. The lintel supporting the masonry consists of three angles, as indicated in Fig. 8-9b. What size should the angles be? *Solution:* (1) A sketch similar to Fig. 8-10 is made to show the dimensions and the triangular load to be supported. By scaling, the height of the triangle is found to be 5.3 ft. Then the wall surface is (6 × 5.3)/2 = 15.9 sq ft—say 16 sq ft. Since the wall is 1 ft thick, the volume of masonry to be supported is 16 × 1 = 16 cu ft. By referring to Table 9-1, we find that common brick weighs 120 lb/cu ft. Hence, the triangular load to be supported is 16 × 120 = 2000 lb, or 2 kips.

(2) The maximum bending moment for this triangular loading (Case 8, Fig. 4-27) is

$$M = \frac{WL}{6} = \frac{2 \times 6 \times 12}{6} = 24 \text{ kip-in.}$$

(3) The required section modulus assuming full lateral support is

$$S = \frac{M}{F_b} = \frac{24}{22} = 1.09 \text{ in.}^3$$

Since three angles are to be used, the section modulus of each angle must be $1.09/3 = 0.36$ in.^{3}

(4) The minimum size angle, a $3\frac{1}{2} \times 3 \times \frac{3}{8}$, with the 3-in. leg vertical has a section modulus of 0.851 in.^{3} (Table 1-5). This section is ample and will be accepted. As a check, it will be of interest to determine just what the value of the bending stress is under this loading. One angle takes a third of the total bending moment or $24/3 = 8$ kip-in. Then

$$f_b = \frac{M}{S} = \frac{8}{0.851} = 9.4 \text{ ksi} < 22 \text{ ksi}$$

This selection also checks out with the rough rule of thumb that calls for the vertical leg to have $\frac{1}{2}$ in. of depth for every foot of the span length.

It should be noted that the arching action of the masonry is not fully attained until the mortar has set. When the mortar is in a plastic state, the loads on the lintel are, of course, somewhat greater than the triangular loading we have computed.

Example 2   An opening in a 12-in. masonry wall is 8 ft 0 in. wide. Conditions are such that the triangular loading may not be assumed, and the total uniform load to be carried has been computed to be 12 kips. Design a lintel composed of the shapes indicated in Fig. 8-9d.

Solution: (1) The maximum bending moment is

$$M = \frac{WL}{8} = \frac{12 \times 8}{8} = 12 \text{ kip-ft}$$

(2) The required section modulus, using the allowable bending stress of 22 ksi for channels laterally supported, is

$$S = \frac{M}{F_b} = \frac{12 \times 12}{22} = 6.55 \text{ in.}^3$$

(3) Assume a make-up for the lintel as indicated below, taking the section moduli from Tables 1-2 and 1-3 for the I-beam and channel, respectively.

$$S \, 5 \times 10 \qquad S = 4.92$$
$$C \, 5 \times 6.7 \qquad S = \underline{3.00}$$
$$\text{Total } S = 7.92 \text{ in.}^3$$

This is acceptable because only 6.55 in.³ is required.

The shelf angle will be an L $3\frac{1}{2} \times 3 \times \frac{5}{16}$ riveted or welded to the channel, with the $3\frac{1}{2}$-in. leg outstanding. The I-beam and channel are held in position by use of one of the types of separators shown in Fig. 8-11. It is not customary to consider the angle as contributing to the strength of the lintel.

Problem 8-12-A. The opening in an 8-in. common brick wall is 5 ft 0 in. wide. Design a lintel composed of angle sections to support the brickwork, assuming there are no additional loads.

Beams 10″ and under

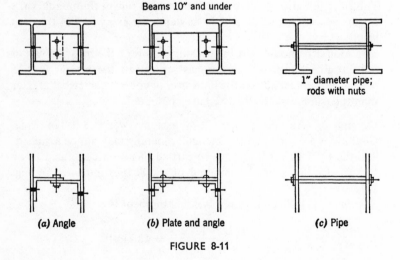

1″ diameter pipe; rods with nuts

(a) Angle   *(b)* Plate and angle   *(c)* Pipe

FIGURE 8-11

*Problem 8-12-B.** A 12-in. brick wall has an opening 10 ft wide to accommodate entrance doorways. The lintel over the opening is made up of an I-beam, channel, and an attached shelf angle similar to the arrangement shown in Fig. 8-9d. If the total uniformly distributed load on the lintel is 21.5 kips, what should be the size of the members?

8-13 Separators

When two beams are used together, such as the I-beam and channel or the two channels of Fig. 8-9d and f, they are held in position by means of separators. These separators tend to give lateral support to the compression flanges and hold the component parts in proper alignment; they are not counted upon to transfer load from one member to the other in event of unequal loading. Three types of separator are shown in Fig. 8-11, the pipe separators being used most commonly for I-beams and channels of shallow depth. In addition to the angle and plate-and-angle separators (Fig. 8-11a and b), a fourth type consists of a single steel plate welded directly to the webs of the component sections. The maximum spacing of separators is often limited to 5 ft, and they should also be placed at the ends of the beams.

8-14 Beam Bearing Plates

Beams that are supported on masonry walls or piers usually rest on steel bearing plates. The purpose of the plate is to provide an ample bearing area. The plate also helps to seat the beam at its proper elevation. Bearing plates provide a level surface for a support and, when properly placed, afford a uniform distribution of the beam reaction over the area of contact with the supporting masonry.

By reference to Fig. 8-12, the area of the bearing plate is $B \times N$. It is found by dividing the beam reaction by F_p, the allowable bearing value of the masonry. Then

$$A = \frac{R}{F_p}$$

in which $A = B \times N$, the area of the plate in sq in.

R = reaction of beam in pounds or kips

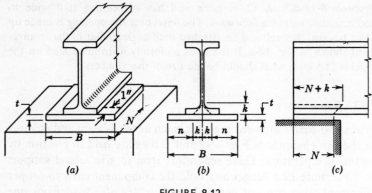

FIGURE 8-12

and F_p = allowable bearing pressure on the masonry in psi or ksi (see Table 8-4).

The thickness of the wall generally determines N, the dimension of the plate parallel to the length of the beam. If the load from the beam is unusually large, the dimension B may become excessive. For such a condition, one or more shallow-depth I-beams, placed parallel to the wall length, may be used instead of a plate. The dimensions B and N are usually in even inches, and a great variety of thicknesses is available.

The thickness of the plate is determined by considering the projection n (Fig. 8-12b) as an inverted cantilever, with the uniform bearing pressure on the bottom of the plate tending to curl it upward about the beam flange. The required thickness may be computed

TABLE 8-4.   Allowable Bearing Pressure on Masonry Walls

Stone concrete, depending on quality	600 or 800 psi
Common brick, lime mortar	100
Common brick, lime-cement mortar	200
Common brick, cement mortar	250
Hard brick, cement mortar	300
Rubble, cement mortar	150
Rubble, lime-cement mortar	100
Hollow T.C. blocks, cement mortar	80
Hollow cinder blocks, cement mortar	80

readily by the formula given below, which does not involve direct computation of bending moment and section modulus.

$$t = \sqrt{\frac{3f_p n^2}{F_b}}$$

in which t = thickness of plate in inches

f_p = *actual* bearing pressure of the plate on the masonry, psi or ksi

F_b = allowable bending stress in the plate. The AISC Specification gives the value of F_b as $0.75F_y$. For A36 steel, $F_y = 36$ ksi; therefore, $F_b = 0.75 \times 36 = 27$ ksi.

$n = \dfrac{B}{2} - k$, in inches (see Fig. 8-12b)

and k = distance from bottom of beam to web toe of fillet in inches. Values of k for various beam sizes may be found in the AISC Manual "Dimensions for detailing" tables.

The above formula is derived by considering a strip of plate 1 in. wide (Fig. 8-12a) and t in. thick, with a projecting length of n inches, as a cantilever. Since the upward pressure on the steel strip is f_p, the bending moment at distance n from the edge of the plate is

$$M = f_p n \times \frac{n}{2} = \frac{f_p n^2}{2}$$

For this strip with rectangular cross section,

$$\frac{I}{c} = \frac{bd^2}{6} \qquad \text{(Art. 5-5)}$$

and since $b = 1$ in. and $d = t$ in.,

$$\frac{I}{c} = \frac{1 \times t^2}{6} = \frac{t^2}{6}$$

Then, from the beam formula,

$$\frac{M}{F_b} = \frac{I}{c} \qquad \text{(Art. 6-1)}$$

Substituting the values of M and I/c determined above,

$$\frac{f_p n^2}{2} \times \frac{1}{F_b} = \frac{t^2}{6}$$

and

$$t^2 = \frac{6 f_p n^2}{2 F_b} \quad \text{or} \quad t = \sqrt{\frac{3 f_p n^2}{F_b}}$$

When determining the dimensions of the bearing plate, the beam should be investigated for web crippling on the length $(N + k)$ shown in Fig. 8-12c. This is explained in Art. 8-15.

Example 1.   A W 21 × 55 of A36 steel transfers a load of 44,000 lb to a wall built of common brick and cement mortar by means of a bearing plate of A36 steel. The dimension N for this plate is 10 in., and the distance k for a W 21 × 55 is 1⅛ in. (1.125 in.). Design a bearing plate for this beam.

Solution: (1) Referring to Table 8-4, we find F_p, the allowable bearing pressure on this type of wall, is 250 psi. The required area of the plate is then

$$A = \frac{R}{F_p} = \frac{44,000}{250} = 176 \text{ sq in.}$$

Since by data $N = 10$ in., $B = 176/10 = 17.6$, say 18 sq in.

(2) This plate is slightly larger than required, so the actual bearing pressure on the wall will be

$$f_p = \frac{R}{A} = \frac{44,000}{18} = 245 \text{ psi}$$

(3) To find the thickness, first determine the value of n:

$$n = \frac{B}{2} - k = \frac{18}{2} - 1.125 = 7.88 \text{ in.}$$

Then

$$t = \sqrt{\frac{3 f_p n^2}{F_b}} = \sqrt{\frac{3 \times 245 \times 7.88 \times 7.88}{27,000}} = 1.3 \text{ in.}$$

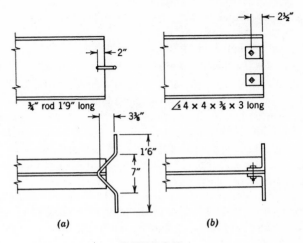

FIGURE 8-13

Working in $\frac{1}{16}$-in. increments for the plate thickness (see Art. 1-7), we accept a plate $10 \times 1\frac{5}{16} \times 1'$-6.

(4) Check for web crippling. The investigation of this beam for web crippling at the support is carried out in Example 1 of Art. 8-15.

Usually the bearing plate is neither riveted nor welded to the beam flange but is shipped separately and grouted in place before erection of the beam. However, beams on masonry walls are usually provided with anchors, such as those shown in Fig. 8-13, as a means of tying the structure together. The type shown in Fig. 8-13a is known as a government anchor, and the bent rod is customarily made $\frac{3}{4}$ in. in diameter. For the angle wall anchor indicated in Fig. 8-13b, the angle sections are usually $4 \times 4 \times \frac{3}{8} \times 3$ in. long, two anchors being used on beams of 12-in. depth and over and one anchor for beams 10 in. deep and under.

In the event a comparatively light reaction or a high allowable bearing pressure reduces the bearing area required so that the beam may be supported without a bearing plate, the beam flange should be checked for bending induced by the bearing pressure to see that it does not exceed F_b. This may be accomplished by use of the formula

$$f_b = \frac{3f_p n^2}{t^2}$$

in which f_p = actual bearing pressure of the beam flange on the masonry, psi or ksi

$\quad\quad\quad\; n$ = (flange width/2) − k, in inches

and $\quad\quad t$ = thickness of *flange*, in inches

Example 2.   A W 12 × 53 of A36 steel transfers a load of 16 kips to a common brick wall laid up with cement mortar. The beam has an 8-in. bearing length (dimension N) on the wall. If a bearing plate is not required, compute the maximum bending stress in the beam flange. The dimension k for a W 12 × 53 is 1¼ in. (1.25 in.). Does the bending stress exceed 27 ksi, the allowable in the flange when acting as a bearing plate?

Solution: (1) Referring to Table 1-1, the flange width (b_f) of a W 12 × 53 is given as 10 in., and the flange thickness (t_f) is given as 0.576 in. From Table 8-4, we find the allowable bearing pressure on the brick wall in 250 psi.

(2) The bearing area of the beam flange on the wall is 8 × 10 = 80 sq in.; and since the reaction is 16 kips, the actual bearing pressure on the wall is

$$f_p = \frac{R}{A} = \frac{16}{80} = 0.200 \text{ ksi or 200 psi}$$

This value is less than the allowable 250 psi so no bearing plate is required.

(3) Investigating the beam flange for bending stress,

$$n = \frac{10}{2} - 1.25 = 3.75 \text{ in.}$$

and

$$f_b = \frac{3f_p n^2}{t^2} = \frac{3 \times 200 \times 3.75 \times 3.75}{0.576 \times 0.576} = 25,400 \text{ psi}$$

which is less than the allowable 27 ksi.

Problem 8-14-A.*   A W 14 × 30 with a reaction of 20 kips rests on a brick wall laid up with common brick and lime-cement mortar. The dimension k for a W 14 × 30 is 1 in., and the beam has a bearing length of 8 in. parallel to the length of the beam. If the bearing plate is A36 steel, determine its dimensions.

Problem 8-14-B.   A wall laid up with common brick and lime-cement mortar supports a W 18 × 50 of A36 steel. The beam reaction is 25 kips and the bearing length N is 9 in. For this beam, the dimension k is $1\frac{1}{16}$ in. Design the beam bearing plate.

Problem 8-14-C. A W 12 × 65 with a reaction of 4 kips is supported by a wall of hard brick laid in cement mortar. The bearing length of the beam on the wall (dimension N) is $3\frac{1}{2}$ in.; the dimension k for the W 12 × 65 is $1\frac{5}{16}$ in.; and the beam is of A36 steel. Determine whether a bearing plate is required. If not, does the bending stress in the beam flange exceed the allowable value?

8-15 Crippling of Beam Webs

An excessive end reaction on a beam or an excessive concentrated load at some point along the interior of the span may cause crippling or buckling of the beam web. The AISC Specification requires that end reactions or concentrated loads for beams without stiffeners or other web reinforcement shall not exceed the following (Fig. 8-14):

$$\text{Maximum end reaction} = 0.75F_y t(N + k)$$
$$\text{Maximum interior load} = 0.75F_y t(N + 2k)$$

in which t = thickness of beam *web*, in inches

N = length of bearing or length of concentrated load (not less than k for end reactions), in inches

k = distance from outer face of flange to web toe of fillet, in inches

and $0.75F_y = 27$ ksi for A36 steel

By substituting 27 ksi for $0.75F_y$ in the foregoing expressions, we can write

$$\text{Maximum end reaction} = 27 \times t \times (N + k)$$
$$\text{Maximum interior load} = 27 \times t \times (N + 2k)$$

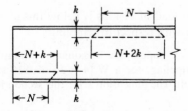

FIGURE 8-14

When these value are exceeded, the webs of the beams should be reinforced with stiffeners, the length of bearing increased, or a beam with thicker web selected.[5]

Example 1.   Carry out the check for web crippling of the beam used in Example 1 of Art. 8-14. The section is a W 21 × 55 of A36 steel, the reaction is 44 kips, and the bearing length $N = 10$ in. The value of k was given as $1\frac{1}{8}$ in. (1.125 in.); and by referring to Table 1-1, we find that the web thickness (t_w) is 0.375 in.

Solution: To investigate for web crippling, we compute the allowable end reaction under the given conditions and compare this with the reaction actually developed by the beam. Then

$$R_{(allowable)} = 27 \times t \times (N + k)$$
$$= 27 \times 0.375 \times (10 + 1.125) = 113 \text{ kips}$$

Since the beam reaction of 44 kips is below this allowable value, the end bearing is safe with respect to web crippling.

Example 2.   A W 12 × 27 of A36 steel supports a column load of 70 kips at the center of its span. The bearing length of the column on the beam is 10 in. The dimension k for a W 12 × 27 is $\frac{15}{16}$ in. (0.9375 in.), and reference to Table 1-1 shows its web thickness (t_w) to be 0.237 in. Investigate this beam for web crippling under the column load.

Solution: The procedure here is to compute the allowable interior load under these conditions and compare this value with the load brought to the beam by the column. Then

$$P_{(allowable)} = 27 \times t \times (N + 2k)$$
$$= 27 \times 0.237 \times [10 + (2 \times 0.9375)] = 76 \text{ kips}$$

Since the column load of 70 kips is below this allowable value, the beam web is safe with respect to crippling.

Problem 8-15-A.   Compute the maximum allowable reaction, with respect to web crippling, for a W 14 × 30 of A36 steel with an 8-in. bearing plate length. The dimension k for this section is 1 in.

---

[5] It should be noted that the values for R, R_i, and N_e, given at the bottom of Table 8-1, are related to the maximum end reaction equation stated above (see Art. 8-7).

Problem 8-15-B.   A column load of 81 kips with a bearing length of 11 in.
is placed on the upper flange of the beam given in the preceding problem.
Are web stiffeners required to prevent web crippling?

8-16   Plate Girders

Although the deeper and heavier wide flange beams fulfill most of
the long span requirements in building construction, plate girders
are employed where unusually heavy loads must be supported over
long spans. Such conditions occur in the lower stories of high-rise

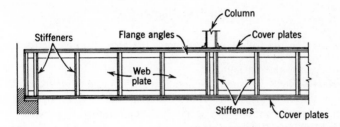

FIGURE 8-15

buildings when certain columns are omitted to increase unobstructed
floor space. This situation requires that one or more of the multistory columns that carry the upper floors be supported on a deep
girder. Figure 8-15 indicates the general arrangement, and Fig. 8-16
shows the makeup of typical plate girder cross sections of both
riveted and welded construction. Welding has largely replaced
riveting in plate girder fabrication, although riveted construction is
permitted by the AISC Specification.

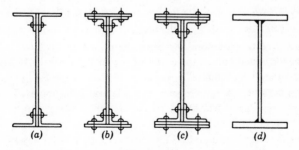

FIGURE 8-16

Design procedures for plate girders are not included in this book. However, the discussion of properties of built-up sections in Art. 5-8 indicates the method of determining gross moment of inertia that forms the basis for proportioning girder cross sections to resist bending. The reader who wishes to pursue the design of plate girders further is referred to the 7th Edition of the AISC Manual. Part 2 of the Manual gives illustrative examples and contains tables to facilitate design under the rules contained in the 1969 AISC Specification.

Review Problems

Note: In the following problems, the structural elements are of A36 steel.

Problem 8-16-A.*   A simple beam adequately braced has a span of 20 ft with a uniformly distributed load, including beam weight, of 400 lb per lin ft extending over its entire length. In addition, there is a concentrated load of 10 kips at 4 ft from one end of the beam. Is a W 12 × 27 adequate for strength in bending? Compute the actual bending stress.

Problem 8-16-B.   If a W 12 × 27 is used for the beam loading given in Problem 8-16-A, investigate the shear.

Problem 8-16-C.*   If the beam section used for the loading in Problem 8-16-A is a W 12 × 27, is the deflection excessive? Compute the actual deflection.

Problem 8-16-D.   A simple beam adequately braced has a span of 24 ft with a uniformly distributed load of 1000 lb per lin ft extending over a distance of 12 ft from one end of the beam. If the beam is a W 10 × 25, investigate the section for strength in bending.

Problem 8-16-E.   Investigate the beam in Problem 8-16-D for shear.

Problem 8-16-F.*   Investigate the beam in Problem 8-16-D for deflection.

Problem 8-16-G.   Suppose the beam described in Problem 8-16-D is laterally supported only at the center of the span. Is a W 10 × 25 acceptable for strength in bending?

Problem 8-16-H.   A simple beam has a laterally unsupported length of 16 ft and a uniformly distributed load of 30 kips. Investigate a W 12 × 36 for strength in bending.

Problem 8-16-I.*   A laterally supported cantilever beam has a length of 8 ft with a uniformly distributed load of 200 lb per lin ft extending over its entire length. In addition, there are two concentrated loads of 1 ton each; one located at the outer end of the beam and the other 4 ft from the support. What is the lightest weight wide flange beam that is adequate for strength in bending?

Problem 8-16-J.   An adequately braced simple beam has a span of 28 ft with a concentrated load of 4 kips located 14 ft from one end. What is the lightest weight standard I-beam that is acceptable with respect to both strength in bending and deflection?

Problem 8-16-K.   A W 16 × 40 has a span of 19 ft and supports a total uniform load of 54 kips. It is supported at its ends on masonry walls made of common brick laid up with cement mortar. Design the bearing plates if the dimension of the plates parallel to the length of the beam is limited to 8 in. For the W 16 × 40, the dimension k is $1\frac{1}{8}$ in.

Problem 8-16-L.*   A wall of common brick laid up with cement mortar supports the end of a W 21 × 62. The beam reaction is 50 kips. Design a bearing plate for this support, assuming that the length of the plate parallel to the length of the beam is 9 in. The dimension k for this beam is $1\frac{1}{4}$ in.

Problem 8-16-M.   A W 16 × 40 has a $3\frac{1}{2}$-in. bearing length at its supports. Compute the maximum allowable reaction for this beam with respect to web crippling. The value of k for this section is $1\frac{1}{8}$ in.

Problem 8-16-N.   A column load of 64 kips at the center of the span on the beam given in Problem 8-16-M has a bearing length of 12 in. Are web stiffeners required to prevent web crippling?

9

Floor
Framing
Systems

||||||||||||||||||||||||||||||||||||||||||||||

9-1 Layout

The arrangement of columns and the layout of beams and girders
depends on a number of factors. The area of the building, the floor
plan, and the occupancy requirements determine the location of
columns and other points of support. When permitted by other
governing conditions, a column spacing of 22 ft, more or less, will
be found economical with respect to the structural work.

The layout of beams and girders depends entirely on the column
spacing and the type of floor system to be employed. Figure 9-1
illustrates two common methods of framing when one of the shorter-
span systems is used, such as the solid concrete slab with wire mesh
reinforcement (Fig. 8-2a) or certain of the corrugated or ribbed steel
deck systems (Fig. 8-2f). Third-point concentration is shown at (a)
in Fig. 9-1 and center-point concentration at (b). When the area
enclosed by the columns (called a *bay*) is not a square, the usual
custom is to run the beams in the long direction, with the girders on

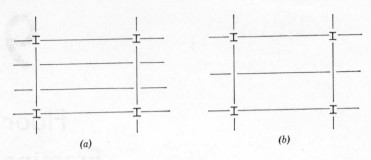

(a)   *(b)*

FIGURE 9-1

the shorter span. This is not always done, however, because the economical length of span of the floor system adopted may be a controlling factor, or it may be necessary that the beam and girder have the same depth to facilitate installation of a flush ceiling. The usual procedure is to design two or more acceptable layouts and then select the most economical.

As stated in Art. 4-3, the area of floor supported by any one beam is equal to the span length multiplied by the sum of half the distances to adjacent beams. The span lengths used in design are generally taken from center to center of supporting members, although this distance may be reduced where beams or girders frame against the flanges of large columns. The total uniform load on a floor beam is then found by multiplying this panel area by the design load in lb per sq ft, the design load for any floor being defined as the sum of the dead and live loads on that floor.

9-2   Dead Load

The dead load consists of the weight of the materials of which the building is constructed: walls, partitions, columns, framing, floors, and roofs. In the design of a beam, the dead load must include an allowance for the weight of the member and its protective covering if fireproofing is required. This weight is first estimated and checked when the design has been completed. Table 9-1, which lists the weights of many construction materials, may be used in the computation of dead loads. The weights given are approximate, and actual weights may vary somewhat from the values given in this table. Most building

TABLE 9-1. Weights of Building Materials

Floors	Pounds per square foot
Board flooring, per inch of thickness	3
Granolithic flooring, per inch of thickness	12
Floor tile, per inch of thickness	10
Asphalt mastic, per inch of thickness	12
Wood block, per inch of thickness	4
Cinder-concrete fill, per inch of thickness	8
Stone-concrete slab, per inch of thickness	12
Slag-concrete slab, per inch of thickness	10
Ceiling, suspended, metal lath and plaster	10
Ceiling, pressed steel	2

Roofs	Pounds per square foot
Three-ply roofing felt and gravel	5½
Five-ply roofing felt and gravel	6½
Roofing tile, cement	15 to 20
Roofing tile, clay, shingle type	12 to 14
Roofing tile, Spanish	8 to 10
Slate, ¼ in. thick	9½
Slate, ⅜ in. thick	12 to 14½
2 in. Book tile	12
Sheathing, wood, 1 in. thick	3
Skylight, ⅜ in. glass in galvanized iron frame	7½

Walls and partitions	Pounds per square foot
8 in. Brick wall	80
12 in. Brick wall	120
17 in. Brick wall	160
4 in. Brick, 8 in. tile backing	75
9 in. Brick, 4 in. tile backing	100
8 in. Wall tile	35
12 in. Wall tile	45
3 in. Clay-tile partition	18
4 in. Clay-tile partition	19
6 in. Clay-tile partition	25
4 in. Glass-block	18
3 in. Gypsum-block partition	11
4 in. Gypsum-block partition	13
2 in. Solid plaster partition	20
4 in. Stud partition, plastered both sides	20
Steel sash, glazed	10

Masonry	Pounds per cubic foot
Ashlar masonry, granite	165
Ashlar masonry, limestone	160
Ashlar masonry, sandstone	140
Brick masonry, common	120
Brick masonry, pressed	140
Concrete, plain stone	145
Concrete, reinforced stone	150
Concrete, cinder	110
Rubble masonry, limestone	150
Rubble masonry, sandstone	130

codes specify the weights of materials to be used in computing dead loads and, of course, these must be used in actual design work. Dead loads are loads due to gravity, and they result in downward vertical forces.

9-3 Live Load

The live load on a floor represents the probable load created by the occupancy. It includes the weight of human occupants, furniture, equipment, stored materials, etc. All building codes give the minimum live loads to be used in the design of buildings for various occupancies. Since there is a lack of uniformity among different building codes in specifying live loads, the local code should always be consulted and followed. Table 9-2 is a compilation of minimum live load requirements assembled from a number of building codes and will serve to show the variation in allowances for different types of occupancy.

Some codes require that floors of manufacturing buildings be designed to support a possible load of 2000 or 3000 lb over an area of 3 or 4 sq ft at any position. Another requirement frequently specified is that 100 psf be taken as the live load for aisles, corridors, lobbies, and public spaces in buildings in which crowds of people are likely to assemble. Although expressed as a uniform load in lb per sq ft, the live load allowances are established large enough to cover the effect of ordinary concentrations which may occur. Where buildings are to contain heavy machinery or similar concentrations of load, these must be provided for individually. The live load on flat roofs must contain an allowance for the weight of snow and ice and for ponding of water.

9-4 Movable Partitions

In office buildings and certain other building types, the partition layout may not be fixed but may be erected or moved from one position to another in accordance with the requirements of the occupant. In order to provide for this flexibility, it is customary to require that an allowance of 15 to 20 psf be added to the design load to cover movable partitions. Since this requirement appears under

TABLE 9-2. Minimum Live Loads

Occupancy or use	Live load, pounds per square foot
Apartments	
Private suites	40
Corridors	100
Rooms for assembly	100
Buildings for public assembly	
Corridors	100
Rooms with fixed seats	60
Rooms with movable seats	100
Dwellings	40
Factories	125
Garages	100
Hotels	
Private rooms	40
Public rooms	100
Office buildings	
Offices	80
Public spaces	100
Restaurants	100
Schools	
Assembly rooms	100
Classrooms with fixed seats	40
Classrooms with movable seats	80
Corridors	100
Stairways and firetowers	100
Stores	
First floors	125
Upper floors	75
Theatres	
Corridors, aisles, and lobbies	100
Fixed seats areas	60
Stage	150

dead loads in some codes and under live loads in others, care must be exercised to determine whether such a provision is mandatory.

9-5 Fireproofing for Beams

In fire resistive construction, some insulating material must be placed around the structural steel to protect it from contact with flames or excessive heat that might cause structural failure. Materials commonly used for this purpose are concrete, masonry, or lath and plaster. In addition, there are fibrous and cementitious coatings that can be sprayed directly on the surfaces of steel members. Figure 8-2a illustrates a method often employed when a reinforced concrete slab is used for the floor system. Regardless of the material, its weight must be given consideration in the design of the beam.

After the beam to be used has been selected, the weight of the fireproofing that projects below the slab may be computed accurately; but in the design of the beam, its probable weight as well as the weight of the fireproofing must be estimated. Building codes differ in the thickness of fireproofing material required. A common specification for beams and girders calls for 2 in. of material on the flat surfaces and $1\frac{1}{2}$ in. on the edges of the flanges. These dimensions are indicated in Fig. 9-2. In this figure, d and b are the depth of beam and width of flange in inches. For the thicknesses shown, the cross-sectional area of the fireproofing (drawn hatched in the figure) is $[d \times (b + 3)]$ sq in. The number of cubic feet of fireproofing *per linear foot of beam* is $[d \times (b + 3)]/144$. Taking the weight of

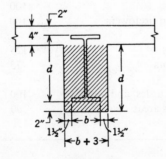

FIGURE 9-2

unreinforced concrete as 144 lb per cu ft, the weight of the fire-proofing per linear foot of beam becomes

$$W_{FP} = \frac{d \times (b + 3)}{144} \times 144 = [d \times (b + 3)] \text{ lb per lin ft}$$

It should be noted that this expression depends upon the thickness of the fireproofing, the thickness of the structural concrete slab, and the distance from top of the steel beam to the top of the slab. Similar equations can, of course, be established for other conditions of encasement. This procedure is sufficiently accurate to make a preliminary allowance for the weight when designing the beam. Table 9-3, based on the above formula, gives the weight of concrete fireproofing for a number of beam sections. When the section has finally been determined, the true weight of the beam and fireproofing are checked to see that an adequate weight allowance is provided.

TABLE 9-3. Approximate Weight of Concrete Fireproofing for Beams, in Pounds per Linear Foot

Wide flange beams		Standard I-beams	
Section	Weight	Section	Weight
W 8 × 17	66	S 6 × 12.5	38
W 10 × 21	86	S 7 × 15.3	47
W 12 × 27	114	S 8 × 18.4	56
W 14 × 30	136	S 10 × 25.4	77
W 16 × 36	159	S 12 × 31.8	96
W 18 × 50	189	S 15 × 42.9	128
W 21 × 62	236	S 18 × 54.7	162
W 24 × 76	287	S 20 × 65.4	185
W 27 × 94	350	S 24 × 79.9	240

9-6 Design of Typical Framing

The following example illustrates the design of beams and girders for a typical bay of floor framing.

Example. The floor plan of a building has columns spaced 20 ft on centers in one direction and 21 ft on centers in the other. The

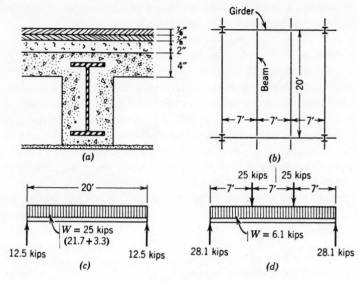

(a)   (b)

(c)   (d)

FIGURE 9-3

floor construction consists of a 4-in. stone concrete slab, 2 in. of
cinder concrete fill, $\frac{7}{8}$-in. wood underflooring, $\frac{7}{8}$-in. wood finished
flooring, and a suspended metal lath and plaster ceiling. The live
load on the floor is 60 psf, and an allowance of 15 psf is to be included
for movable partitions. Design the framing of A36 steel for third-
point concentration with the beams spanning the shorter dimension.
Solution for Beam Design: (1) Make sketches (Fig. 9-3a, b, and c)
showing the construction, framing layout, and beam loading. The
design load is computed with the aid of Table 9-1:

Dead load:			
4-in. concrete slab	=	48	psf
2-in. cinder concrete fill	=	16	
Underflooring	=	3	
Finished flooring	=	3	
Suspended ceiling	=	10	
Movable partitions	=	15	
Live load	=	60	
Total design load	=	155	psf

(2) The *superimposed* uniform load on one beam is the panel area multiplied by the design load, or

$$W = 7 \times 20 \times 155 = 21.7 \text{ kips}$$

Computation and recording of reactions will be deferred until the weight of beam and fireproofing have been determined.

(3) The maximum bending moment due to the superimposed load is

$$M = \frac{WL}{8} = \frac{21.7 \times 20}{8} = 54.3 \text{ kip-ft}$$

(4) Since encasement of the top flange of the beam in the concrete slab provides full lateral support, the allowable bending stress is 24 ksi (Table 3-2), and the required section modulus is

$$S = \frac{M}{F_b} = \frac{54.3 \times 12}{24} = 27.15 \text{ in.}^3$$

(5) Referring to Table 8-2, it is found that the nearest section modulus tabulated is 29.4 in.^{3} for an S 10 × 35. Since we have not yet taken the weight of the beam and its fireproofing into account, a beam with a larger section modulus should be selected for trial. Therefore, try a W 14 × 30 with a section modulus of 41.9 in.^{3}

From Table 9-3, the approximate weight of fireproofing for this section is 136 lb per lin ft, making total beam weight $(30 + 136) \times 20 = 3.32$ kips. The revised total load is therefore $21.7 + 3.32 = 25.0$ kips, and the revised maximum bending moment becomes

$$M = \frac{WL}{8} = \frac{25 \times 20}{8} = 62.5 \text{ kip-ft}$$

The section modulus required by this revised bending moment is

$$S = \frac{M}{F_b} = \frac{62.5 \times 12}{24} = 31.25 \text{ in.}^3$$

Table 8-2 shows that a W 12 × 27 has an S of 34.2 in.^{3} and that it will support a uniform load of 27.4 kips on a 20-ft span. However, the table also shows that the deflection would be excessive under this load, and we may judge that it probably would be excessive also under the 25 kips to be carried. Scanning Table 8-2 further, it is seen that a W 14 × 30 ($S = 41.9$) will support 33 kips on a 20-ft span and consequently might carry 25 kips without excessive deflection. We accept a W 14 × 30 subject to checking for shear and deflection.

(6) Testing for shear, we note that the beam reactions under the revised load of 25 kips are each 12.5 kips, and we find from Table 1-1 that the depth and web thickness of this section are: $d = 13.86$ in. and $t_w = 0.27$ in. Then

$$f_v = \frac{V}{A_w} = \frac{12.5}{13.86 \times 0.27} = 3.34 \text{ ksi}$$

Since the allowable shearing stress is 14.5 ksi (Table 3-2), this beam is acceptable for shear.

(7) The allowable deflection is (20 × 12)/360 or 0.67 in., and the actual deflection is given by Case 2, Fig. 4-27 as

$$\Delta = \frac{5Wl^3}{384EI}$$

The moment of inertia of a W 14 × 30 is given in Table 1-1 as 290 in.[4] Then

$$\Delta = \frac{5 \times 25 \times (20 \times 12)^3}{384 \times 29{,}000 \times 290} = 0.536 \text{ in.}$$

This does not exceed the allowable 0.67 in., so the W 14 × 30 is accepted. The final reactions of 12.5 kips should now be recorded on the load diagram of Fig. 9-3c.

Solution for Girder Design: (1) Referring to Fig. 9-3b, we note that the girders receive the reactions of *two* typical floor beams at each third-point, making the value of the concentrated loads 2 × 12.5 = 25 kips. These are recorded on the load diagram of Fig. 9-3d and constitute the superimposed load on a girder.

(2) Recording of reactions is deferred until those due to the full load, including girder and fireproofing weight, have been determined.

(3) The maximum bending moment due to the superimposed load (Case 3, Fig. 4-27) is

$$M = \frac{PL}{3} = \frac{25 \times 21}{3} = 175 \text{ kip-ft}$$

(4) Since the top flange of the girder is fully encased in the concrete slab in addition to being braced at the third-points, the allowable bending stress is 24 ksi and the required section modulus is

$$S = \frac{M}{F_b} = \frac{175 \times 12}{24} = 87.5 \text{ in.}^3$$

(5) Referring to Table 8-2, it is seen that a W 18 × 50 has an S of 89.1 in.^{3}. However, the margin of 1.6 in.^{3} provided by this section will not be sufficient to cover the additional S required by the weight of the beam and fireproofing. As a trial section, we might select a W 18 × 55 ($S = 98.4$); but in view of the span length and loading, this section might have excessive deflection. Therefore, we will select a W 21 × 55 for trial ($S = 110$).

From Table 9-3, the approximate weight of fireproofing for this beam is 236 lb per lin ft (listed for a W 21 × 62, which has approximately the same depth and flange width). This results in a total beam weight of $(55 + 236) \times 21 = 6.1$ kips and a maximum bending moment at the center of the span of

$$M = \frac{WL}{8} = \frac{6.1 \times 21}{8} = 16 \text{ kip-ft}$$

The section modulus required by this bending moment is

$$S = \frac{M}{F_b} = \frac{16 \times 12}{24} = 8.0 \text{ in.}^3$$

and the total section modulus required is $87.5 + 8.0 = 95.5$ in.^{3} The W 21 × 55 with its S of 110 in.^{3} is more than adequate and is adopted subject to checking for shear and deflection.

(6) Testing for shear, V is equal to the girder reaction, 25 + 6.1/2 or 28.1 kips. The depth and web thickness of the W 21 × 55 are given in Table 1-1 as $d = 20.80$ in. and $t_w = 0.375$ in. Then

$$f_v = \frac{V}{A_w} = \frac{28.1}{20.80 \times 0.375} = 3.6 \text{ ksi}$$

This value is less than the allowable 14.5 ksi, so the W 21 × 55 is acceptable for shear.

(7) If deflection is limited to $\frac{1}{360}$ of the span for total dead and live load, the allowable $\Delta = (21 \times 12)/360 = 0.7$ in. Because the

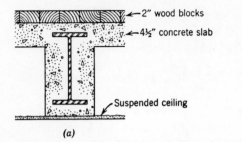

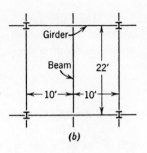

(a)   (b)

FIGURE 9-4

maximum deflection from the third-point loading and that from the uniform load occur at the center of the span, the maximum deflection of the beam will be the sum of these deflections computed separately (Example 2, Art. 7-4). The moment of inertia of a W 21 × 55 is found in Table 1-1 to be 1140 in.[4], and for third-point loading (Case 3, Fig. 4-27) the deflection is

$$\Delta = \frac{23Pl^3}{648EI} = \frac{23 \times 25 \times (21 \times 12)^3}{648 \times 29,000 \times 1140} = 0.429 \text{ in.}$$

For the uniform load (Case 2, Fig. 4-27), the deflection is

$$\Delta = \frac{5Wl^3}{384EI} = \frac{5 \times 6.1 \times (21 \times 12)^3}{384 \times 29,000 \times 1140} = 0.038 \text{ in.}$$

The total deflection is 0.429 + 0.038 = 0.467 in.; and since this is less than the allowable 0.7 in., the W 21 × 55 is accepted for the girder.

The loading diagram of Fig. 9-3d is now completed by recording the final girder reactions of 28.1 kips each.

Problem 9-6-A. The floor framing for a typical interior bay of a building is shown in Fig. 9-4b, and a detail of the floor construction is shown in Fig. 9-4a. Design the beams and girders for a live load of 100 psf, using A36 steel.

9-7 Open Web Steel Joists

Open web steel joists are shop-fabricated, parallel chord trusses used in buildings for the direct support of floors and roof decks

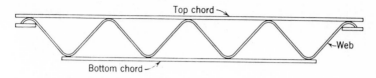

FIGURE 9-5

between main supporting beams or walls (Figs. 9-5 and 8-2c). Floors and roof decks may consist of cast-in-place or precast concrete or gypsum, formed steel, or wood. The usual maximum spacing of joists is 24 in., but the spacing must not exceed the safe span of the wood deck or other material that is used over them.

When a wood deck is placed over the steel joists, it should be fastened to the joists by wood nailing strips (nailers) attached to the top chords of the joists. Cast-in-place concrete slabs reinforced with ribbed metal lath should not be less than 2 in. thick. The attachment for the slab or deck to the top chords of the joists must be so detailed as to brace the top chord against lateral buckling.

Open web joists are manufactured in six different series, but the discussion here will be limited to the J-Series, which is fabricated from A36 steel and is assigned an allowable bending stress of 22,000 psi by the Steel Joist Institute's 1965 *Standard Specifications for Open Web Steel Joists J-Series and H-Series*.[1] Information concerning the makeup and properties of specific joist types, safe load tables, and construction details are contained in the catalogs of the various manufacturers. (See also Sweet's Architectural Catalog File.)

Table 9-4 lists the safe uniformly distributed loads in pounds per linear foot of the J-Series joists. To identify the joists, the numerals that precede the letters give the nominal depth of the joist in inches, the letter or letters indicate the series, and the numerals on the right designate the chord sections. The J-Series joists are made in standard

---

[1] The chord sections of H-Series joists are fabricated from steel having a yield point of 50,000 psi, and a bending stress of 30,000 psi is permitted. The LJ- and LH-Series cover Longspan Steel Joists, and the DLJ- and DLH-Series denote the Deep Longspan Steel Joists intended for use in roof construction. Specifications for all of these series, as well as load tables, are contained in Part 5 of the AISC Manual, 7th Edition.

TABLE 9-4. Standard Load Table for Open Web Steel Joists, J-Series,† Allowable Total Safe Loads in Pounds per Linear Foot, Based on Allowable Stress of 22,000 psi

Joist designation	8J3	10J3	10J4	12J3	12J4	12J5	12J6	14J3	14J4	14J5	14J6	14J7	16J4	16J5	16J6	16J7	16J8
*Depth in inches	8	10	10	12	12	12	12	14	14	14	14	14	16	16	16	16	16
Resisting moment in inch kips	70	89	111	108	135	161	196	127	159	190	230	276	173	216	258	310	359
Max. end reaction in pounds	2000	2200	2400	2300	2500	2700	3000	2400	2800	3100	3400	3700	3000	3300	3600	4000	4300
** Approx. joist wgt. pounds per foot	4.8	4.8	6.0	5.1	6.0	7.0	8.1	5.2	6.4	7.3	8.4	9.7	6.6	7.6	8.5	10.1	11.3
Span in feet																	
8	400																
9	364	440	480														
10	324	400	436														
11	276	367	400	383													
12	238	338	369	354	417	450	500										
13	207	303	343	329	385	415	462	343	400	443	486	529					
14	182	264	320	307	357	386	429	320	373	413	453	493					
15	161	232	289	281	313	360	400	300	350	388	425	463	375	413	450	500	538
16	144	205	256	249	294	338	375	282	329	365	400	435	353	388	424	471	506
17	129	183	228	222	278	318	353	261	311	344	378	411	333	367	400	444	478
18	117	164	205	199	249	300	333	235	294	326	358	389	316	347	379	421	453
19		148	185	180	225	284	316	212	265	310	340	370	288	330	360	400	430
20				163	204	268	300	192	240	287	324	352	262	314	343	381	410
21				149	186	243	286	175	219	262	309	336	238	298	327	364	391
22				136	170	222	270	160	200	239	290	322	218	272	313	348	374
23				125	156	203	247	147	184	220	266	308	200	250	299	333	358
24						186	227	135	170	203	245	294	185	230	275	320	344
25								125	157	187	227	272	171	213	254	306	331
26								116	145	174	210	252	158	198	236	283	319
27								108	135	162	196	235	147	184	219	264	305
28													137	171	205	246	285
29													128	160	191	230	266
30													120	150	179	215	249
31													113	141	168	202	234
32																	

Loads above heavy stepped lines are governed by shear.
† Copyright 1965, Steel Joist Institute. Reprinted by permission.
* Indicates nominal depth of steel joists only.
** Approximate weights per linear foot of steel joists only. Accessories and nailer strip not included.

on Allowable Stress of 22,000 psi (Continued)

Joist designation	1815	1816	18J7	1818	2015	2016	2017	7018	2216	2217	2218	2416	2417	2418
* Depth in inches	18	18	18	18	20	20	20	20	22	22	22	24	24	24
Resisting moment in inch kips	243	293	352	406	265	316	382	455	335	420	493	367	460	540
Max. end reaction in pounds	3500	3900	4200	4500	3800	4100	4300	4600	4200	4500	4800	4400	4700	5000
** Approx. joist wgt. pounds per foot	7.9	9.0	10.2	11.3	8.1	9.2	10.6	11.9	9.6	10.5	11.9	9.9	11.1	12.4
Span in feet														
16														
17														
18	389	433	467	500										
19	368	411	442	474										
20	350	390	420	450	380	410	430	460						
21	333	371	400	429	362	390	410	438						
22	318	355	382	409	345	373	391	418	382	409	436			
23	304	339	365	391	330	357	374	400	365	391	417			
24	281	325	350	375	307	342	358	383	350	375	400	367	392	417
25	259	312	336	360	283	328	344	368	336	360	384	352	376	400
26	240	289	323	346	261	312	331	354	323	346	369	338	362	385
27	222	268	311	333	242	289	319	341	306	333	356	326	348	370
28	207	249	299	321	225	269	307	329	285	321	343	312	336	357
29	193	232	279	310	210	250	297	317	266	310	331	291	324	345
30	180	217	261	300	196	234	283	307	248	300	320	272	313	333
31	169	203	244	282	184	219	265	297	232	290	310	255	303	323
32	158	191	229	264	173	206	249	288	218	273	300	239	294	313
33	149	179	215	249	162	193	234	279	205	257	291	225	282	303
34	140	169	203	234	153	182	220	262	193	242	282	212	265	294
35	132	159	192	221	144	172	208	248	182	229	268	200	250	286
36	125	151	181	209	136	163	197	234	172	216	254	189	237	278
37					129	154	186	222	163	205	240	179	224	263
38					122	146	176	210	155	194	228	169	212	249
39					116	139	167	199	147	184	216	161	202	237
40					110	132	159	190	140	175	205	153	192	225
41									133	167	196	146	182	214
42									127	159	186	139	174	204
43									121	151	178	132	166	195
44									115	145	170	126	158	186
45												121	151	178
46												116	145	176
47												111	139	163
48												106	133	156

* Indicates nominal depth of steel joists only.
** Approximate weights per linear foot of steel joists only. Accessories and nailer strip not included.

depths of 8 to 24 in. in 2-inch increments, and their lengths accommodate spans up to 48 ft. The loads tabulated in Table 9-4 are total loads; to find the *live* loads, the *dead* loads, including the weight of the joists, must be deducted.

The following example illustrates design procedures that may be used in determining the size and spacing of open web steel joists.

Example 1.   Open web steel joists are to support a floor on which the live load is 60 psf. The joist span is 18 ft, and the construction consists of a 2-in. reinforced concrete slab, rough and finish board flooring, and a suspended metal lath and plaster ceiling. Determine the size and spacing of the joists.

Solution: (1) Compute the design load on the floor, using Table 9-1 for weights of materials.

> *Dead load:*
>
> 2-in. Reinforced concrete slab = 24 psf
> 2-in. Board flooring   = 6
> Suspended plaster ceiling   = 10
> *Live load*   = 60
> *Total superimposed load*   = 100 psf

(2) Since the 2-in. concrete slab can be used with a joist spacing of 24 in., we accept this spacing. With a 24-in. spacing, each linear foot of joist supports 2 sq ft of floor area. Consequently, each linear foot of joist supports a load of $100 \times 2 = 200$ lb, exclusive of the weight of the joist. Scanning the loads tabulated for 18 ft in Table 9-4, it will be seen that a joist weighing 5 or 6 lb per lin ft will probably have to be used. Therefore, the total load becomes $200 + 6 = 206$ lb per lin ft of joist.

(3) Referring again to Table 9-4, we see that the 228-lb load tabulated for a 10J4 exceeds the 206-lb load to be carried, as does the 222-lb load listed for a 12J3. Therefore, either of these joists will be acceptable. If the additional 2 inches of depth of the 12J3 is not objectionable from the standpoint of ceiling height in the story below, this joist, which weighs 5.1 lb per lin ft, would be adopted.

(4) Checking the end reaction, the allowable value for a 12J3 is listed in Table 9-4 as 23,000 lb. The reaction developed by the load is $(206 \times 18)/2 = 1854$ lb, which is well within the allowable value.

Alternate Solution: Since Table 9-4 gives the resisting moment for each open web joist, another method for selecting the proper joist size is to compute the maximum bending moment. Using the information developed in Step 2 above, the total uniformly distributed load on one joist is 206 × 18 = 3708 lb—say 3.71 kips. Each joist acts as a simple beam, so the maximum bending moment is

$$M = \frac{WL}{8} = \frac{3.71 \times 18}{8} = 8.35 \text{ kip-ft or } 100 \text{ kip-in.}$$

From Table 9-4, we find that a resisting moment of 111 kip-in. is provided by a 10J4 and that a resisting moment of 108 kip-in. is provided by a 12J3. Since the bending moment developed by the loading is less than either of these values, either a 10J4 or a 12J3 is acceptable.

Example 2.   For the same floor load and span given in Example 1, determine the size of open web steel joists required if their spacing is 16 in. on centers.

Solution: With a 16-in. spacing, each linear foot of joist supports 16/12 or 1.33 sq ft of floor area, making the superimposed load 1.33 × 100 = 133 lb per lin ft of joist. Scanning Table 9-4, we see that an 8J3 weighing 4.8 lb per lin ft will support a total load of 144 lb per lin ft on the 18-ft span. The total load then becomes 133 + 4.8 = 138 lb per lin ft, which is less than the allowable load, so the 8J3 is acceptable.

Problem 9-7-A.   Determine the size of open web steel joists for a span of 21 ft and a live load of 50 psf. The joist spacing is 24 in. on centers, and the floor construction consists of a 2-in. reinforced concrete slab on which is placed rough and finished wood flooring. A plaster ceiling on metal lath is suspended from the bottom chords of the joists.

Problem 9-7-B.   For the data given in Problem 9-7-A, determine the size of open web steel joists if the spacing is made 18 in. on centers.

Alternate Solution. Since Table 9-4 gives the resisting moment for conventional steel-pole, another method for selecting the proper joist size is to compute the maximum bending moment. Using this information developed in Step 2 above, the total uniformly distributed load per lineal foot is $2000 \times 16 = 3700$ lb. Since 3.11 sips each joist acts as a simple beam, so the maximum bending moment is

$$M = \frac{WL}{8} = \frac{3.11 \times 16}{8} = 6.35 \text{ kip-ft or 1100 in-lb}$$

From Table 9-4, we find that a resisting moment of 111 kip-ft, approached by a 1014 and that a resisting moment of 108 kip-ft, is superseded even 12 is since the bending moment developed by the loading is less than either of these joists, either 8-1014 or a 121 is acceptable.

Example 2. For the same download and span given in Example 1, determine the joist when web steel joists required if their spacing is 16 in. on centers.

Solution. When joists spacing each linear foot of floor supports $16,000 \times 1.33 = 21,300$ lb/linear foot. Using the superimposed load $1.33 \times 100 = 1.72$ lb, let us look up axes. Scanning Table 9-4, we see that 1014 needing 4.5 lb per lb-ft will support a total load of 115 + 1.66 = 217 lb per lin ft open-web joist load, then because 115 + 1.66 = 217 lb per lin ft, this is less than the allowable load of the 815 is acceptable.

Problem 9-7-A. Determine the steel open-web steel joists for a span of 21 ft and live loading 50 psf. The joist spacing is 18-in. on centers and the floor construction consists of a reinforced concrete slab on when is placed rough and finished wood flooring. A plaster ceiling on metal lath is supported from the bottom chord of the joist.

Problem 9-7-B. For the data given in Problem 9-7-A, determine the steel open-web steel joists if the span is 18 ft on centers.

10

Columns

||||||||||||||||||||||||||||||||||||||||||||||||||||

10-1 Introduction

A column or strut is a compression member, the length of which is
several times greater than its least lateral dimension. The term
column denotes a relatively heavy vertical member, whereas the
lighter vertical and inclined members, such as braces and the
compression members of roof trusses, are called struts.

Under the discussion of direct stress in Art. 2-2, it was pointed
out that the unit compressive stress in the short block shown in
Fig. 2-1b could be expressed by the direct stress formula $f_a = P/A$;
but it was also stated that this relationship became invalid as the
ratio of the length of the compression member to its least width
increased. To pursue this further, consider a small block of steel
1 in. by 1 in. in cross section and 1 or 2 inches high. If the allowable
compressive stress is 20 ksi, the block will safely support a load P
(Fig. 10-1a) of 20 kips. If, however, we consider a bar of the same
cross section with a length of, say, 20 to 30 in., we find that the value
of P that it will sustain is considerably less because of the tendency
of this more slender bar to buckle or bend (Fig. 10-1b). Therefore,
in columns, the element of *slenderness* must be taken into account
when determining allowable loads. A short column or block fails
by crushing, but long slender columns fail by stresses that result
from bending.

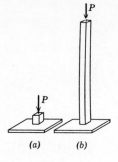

FIGURE 10-1

10-2 Column Shapes

Because of the tendency to bend, the safe load on a column depends not only on the number of square inches in the cross section but also on the manner in which the material is distributed with respect to the axes of the cross section; that is, the *shape* of the column section is an important factor. An axially loaded column tends to bend in a plane perpendicular to the axis of the cross section about which the moment of inertia is least. Since column cross sections are seldom symmetrical with respect to both major axes, the ideal section would be one in which the moment of inertia for each major axis is equal. Pipe columns and structural tubing meet this condition, but their use is somewhat limited because of difficulties in making beam connections.

Of the two major axes of a standard I-beam, the moment of inertia about the axis parallel to the web is much the smaller; hence, for the amount of material in the cross section, I-beams are not economical shapes when used as columns or struts. In former years, built-up sections such as Fig. 10-2c and d were used extensively, but wide flange sections (Fig. 10-2a) are now rolled in a large variety of sizes and are used universally because they require a minimum of fabrication. They are sometimes called H-columns. For excessive loads or unusual conditions, plates are riveted or welded to the flanges of wide flange sections to give added strength (Fig. 10-2b). The compression members of steel trusses are usually formed of two angles, as shown in Fig. 10-2e.

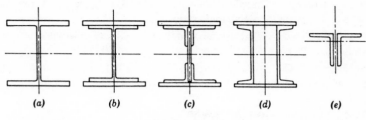

(a)   (b)   (c)   (d)   (e)

FIGURE 10-2

10-3 Slenderness Ratio

In the design of timber columns, the term *slenderness ratio* is defined as the unbraced length divided by the dimension of the least side, both in inches. For structural shapes such as those shown in Fig. 10-2, the least lateral dimension is not an accurate criterion; and the radius of gyration r, which relates more precisely to the stiffness of columns in general, is used in steel column design. As discussed in Art. 5-7, $r = \sqrt{I/A}$. For rolled sections, the value of the radius of gyration with respect to both major axes is given in the tables of properties for designing. For built-up sections, it may be necessary to compute its value. The slenderness ratio of a steel column is then l/r, in which l is the effective length of the column in inches and r is the *least* radius of gyration of the cross section, also in inches. The slenderness ratio for main compression members should not exceed 200.

Example. A W 10 × 49 is used as a column whose effective length is 20 ft. Compute the slenderness ratio.

Solution: Referring to Table 1-1, the radii of gyration for this cross section are $r_{X-X} = 4.35$ in. and $r_{Y-Y} = 2.54$ in. Therefore, the *least* radius of gyration is 2.54 in.

Since the effective length of the column is 20 ft, the slenderness ratio is

$$\frac{l}{r} = \frac{20 \times 12}{2.54} = 94.4$$

It should be remembered that the tendency to bend under the axial loading is in a plane perpendicular to the axis about which the radius of gyration is least.

Problem 10-3-A. The effective length of a W 8 × 31 used as a column is 16 ft. Compute the slenderness ratio.

Problem 10-3-B. What is the slenderness ratio of a column whose section is a W 12 × 65 with an effective length of 30 ft?

Problem 10-3-C.* A single angle 6 × 4 × ½ is to be used as a strut. If the length is 8 ft 6 in. compute the slenderness ratio.

10-4 Effective Column Length

The AISC Specification requires that, in addition to the unbraced length of a column, the condition of the ends must be given consideration. The slenderness ratio is given as Kl/r, in which K is a factor dependent on the restraint at the ends of a column and the means available to resist lateral motion. Figure 10-3 shows diagrammatically six idealized conditions in which joint rotation and

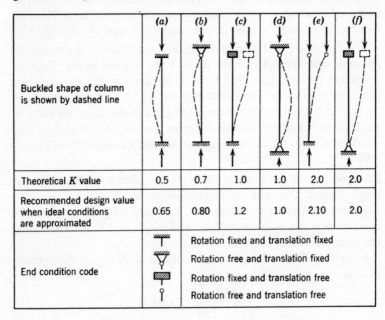

	(a)	(b)	(c)	(d)	(e)	(f)
Buckled shape of column is shown by dashed line						
Theoretical K value	0.5	0.7	1.0	1.0	2.0	2.0
Recommended design value when ideal conditions are approximated	0.65	0.80	1.2	1.0	2.10	2.0
End condition code	Rotation fixed and translation fixed Rotation free and translation fixed Rotation fixed and translation free Rotation free and translation free					

FIGURE 10-3. Effect of end conditions. Reproduced from the 7th Edition of the *Manual of Steel Construction*. Courtesy American Institute of Steel Construction.

joint translation are illustrated. The term K is the ratio of the effective column length to the actual unbraced length. For average conditions in building construction the value of K is taken as 1 so that the slenderness ratio Kl/r becomes simply l/r. See Fig. 10-3d.

10-5 Column Formulas

The AISC Specification gives the following requirements for use in the design of compression members. The allowable unit stresses shall not exceed the following values:

On the gross section of axially loaded compression members, when Kl/r, the largest effective slenderness ratio of any unbraced segment is less than C_c,

$$F_a = \frac{\left[1 - \dfrac{(Kl/r)^2}{2C_c^2}\right]F_y}{\text{F.S.}} \qquad \text{(Formula 10-5-1)}$$

where

$$\text{F.S.} = \text{factor of safety} = \frac{5}{3} + \frac{3(Kl/r)}{8C_c} - \frac{(Kl/r)^3}{8C_c^3}$$

and

$$C_c = \sqrt{\frac{2\pi^2 E}{F_y}}$$

On the gross section of axially loaded columns, when Kl/r exceeds C_c,

$$F_a = \frac{12\pi^2 E}{23(Kl/r)^2} \qquad \text{(Formula 10-5-2)}$$

On the gross section of axially loaded bracing and secondary members, when l/r exceeds 120 (for this case, K is taken as unity),

$$F_{as} = \frac{F_a \text{ (by Formula 10-5-1 or 10-5-2)}}{1.6 - \dfrac{l}{200r}} \qquad \text{(Formula 10-5-3)}$$

In these formulas,

F_a = the axial compression stress permitted in the absence of bending stress, in psi or ksi

K = effective length factor (see Art. 10-4)

l = actual unbraced length, in inches

r = governing radius of gyration (usually the least), in inches

$$C_c = \sqrt{\frac{2\pi^2 E}{F_y}}; \text{ for A36 steel, } C_c = 126.1$$

F_y = the minimum yield point of the steel being used (for A36 steel, $F_y = 36,000$), in psi

F.S. = factor of safety (see above)

E = the modulus of elasticity of structural steel, 29,000,000 psi

F_{as} = the axial compressive stress, permitted in the absence of bending stress, for bracing and other secondary members, in psi or ksi

To determine the allowable axial load that a main column will support, F_a, the allowable unit stress, is computed by formula 1 or 2, and this stress is multiplied by the cross-sectional area of the column. If the column is a secondary member or is used for bracing, formula 3 gives the allowable unit stress; these allowable unit stresses are somewhat greater than those permitted for main members. Table 10-1 gives allowable stresses computed in accordance with these formulas. It should be examined carefully since it will be found to be of great assistance. Note particularly that this table is for use with A36 steel; tables based on other grades of steel are contained in the AISC Manual.

10-6 Allowable Column Loads

The allowable axial load that a steel column will support is found by multiplying the allowable unit stress by the cross-sectional area of the column. The value of Kl/r is first determined and, by referring to Table 10-1, we can establish the allowable unit stress.

Example 1. A W 12 × 65 is to be used as a main column in a building. Its unbraced height is 16 ft. Compute the allowable load on the column.

Solution: (1) Referring to Table 1-1, we find that the cross-sectional area of a W 12 × 65 is 19.1 sq in., $r_{X-X} = 5.28$ in., and $r_{Y-Y} = 3.02$ in. Since the column is unbraced with respect to both major

TABLE 10-1. Allowable Unit Stresses for Columns of A36 Steel,* in kips per square inch

Main and secondary members Kl/r not over 120						Main members Kl/r 121 to 200				Secondary members† l/r 121 to 200			
$\frac{Kl}{r}$	Fa (ksi)	$\frac{Kl}{r}$	Fa (ksi)	$\frac{Kl}{r}$	Fa (ksi)	$\frac{Kl}{r}$	Fa (ksi)	$\frac{Kl}{r}$	Fa (ksi)	$\frac{l}{r}$	Fas (ksi)	$\frac{l}{r}$	Fas (ksi)
1	21.56	41	19.11	81	15.24	121	10.14	161	5.76	121	10.19	161	7.25
2	21.52	42	19.03	82	15.13	122	9.99	162	5.69	122	10.09	162	7.20
3	21.48	43	18.95	83	15.02	123	9.85	163	5.62	123	10.00	163	7.16
4	21.44	44	18.86	84	14.90	124	9.70	164	5.55	124	9.90	164	7.12
5	21.39	45	18.78	85	14.79	125	9.55	165	5.49	125	9.80	165	7.08
6	21.35	46	18.70	86	14.67	126	9.41	166	5.42	126	9.70	166	7.04
7	21.30	47	18.61	87	14.56	127	9.26	167	5.35	127	9.59	167	7.00
8	21.25	48	18.53	88	14.44	128	9.11	168	5.29	128	9.49	168	6.96
9	21.21	49	18.44	89	14.32	129	8.97	169	5.23	129	9.40	169	6.93
10	21.16	50	18.35	90	14.20	130	8.84	170	5.17	130	9.30	170	6.89
11	21.10	51	18.26	91	14.09	131	8.70	171	5.11	131	9.21	171	6.85
12	21.05	52	18.17	92	13.97	132	8.57	172	5.05	132	9.12	172	6.82
13	21.00	53	18.08	93	13.84	133	8.44	173	4.99	133	9.03	173	6.79
14	20.95	54	17.99	94	13.72	134	8.32	174	4.93	134	8.94	174	6.76
15	20.89	55	17.90	95	13.60	135	8.19	175	4.88	135	8.86	175	6.73
16	20.83	56	17.81	96	13.48	136	8.07	176	4.82	136	8.78	176	6.70
17	20.78	57	17.71	97	13.35	137	7.96	177	4.77	137	8.70	177	6.67
18	20.72	58	17.62	98	13.23	138	7.84	178	4.71	138	8.62	178	6.64
19	20.66	59	17.53	99	13.10	139	7.73	179	4.66	139	8.54	179	6.61
20	20.60	60	17.43	100	12.98	140	7.62	180	4.61	140	8.47	180	6.58
21	20.54	61	17.33	101	12.85	141	7.51	181	4.56	141	8.39	181	6.56
22	20.48	62	17.24	102	12.72	142	7.41	182	4.51	142	8.32	182	6.53
23	20.41	63	17.14	103	12.59	143	7.30	183	4.46	143	8.25	183	6.51
24	20.35	64	17.04	104	12.47	144	7.20	184	4.41	144	8.18	184	6.49
25	20.28	65	16.94	105	12.33	145	7.10	185	4.36	145	8.12	185	6.46
26	20.22	66	16.84	106	12.20	146	7.01	186	4.32	146	8.05	186	6.44
27	20.15	67	16.74	107	12.07	147	6.91	187	4.27	147	7.99	187	6.42
28	20.08	68	16.64	108	11.94	148	6.82	188	4.23	148	7.93	188	6.40
29	20.01	69	16.53	109	11.81	149	6.73	189	4.18	149	7.87	189	6.38
30	19.94	70	16.43	110	11.67	150	6.64	190	4.14	150	7.81	190	6.36
31	19.87	71	16.33	111	11.54	151	6.55	191	4.09	151	7.75	191	6.35
32	19.80	72	16.22	112	11.40	152	6.46	192	4.05	152	7.69	192	6.33
33	19.73	73	16.12	113	11.26	153	6.38	193	4.01	153	7.64	193	6.31
34	19.65	74	16.01	114	11.13	154	6.30	194	3.97	154	7.59	194	6.30
35	19.58	75	15.90	115	10.99	155	6.22	195	3.93	155	7.53	195	6.28
36	19.50	76	15.79	116	10.85	156	6.14	196	3.89	156	7.48	196	6.27
37	19.42	77	15.69	117	10.71	157	6.06	197	3.85	157	7.43	197	6.26
38	19.35	78	15.58	118	10.57	158	5.98	198	3.81	158	7.39	198	6.24
39	19.27	79	15.47	119	10.43	159	5.91	199	3.77	159	7.34	199	6.23
40	19.19	80	15.36	120	10.28	160	5.83	200	3.73	160	7.29	200	6.22

Note: $C_c = 126.1$

* Reproduced from *Manual of Steel Construction*. Courtesy American Institute of Steel Construction.
† K taken as 1.0 for secondary members.

axes, the least radius of gyration will be used to determine the slenderness ratio. Also, as in the average end conditions in building construction (Fig. 10-3d), $K = 1$. The slenderness ratio is then

$$\frac{Kl}{r} = 1 \times \frac{16 \times 12}{3.02} = 63.6$$

(2) To determine the allowable unit compressive stress, we turn to Table 10-1. Here we find that, for $Kl/r = 63$, $F_a = 17.14$ ksi and, for $Kl/r = 64$, $F_a = 17.04$ ksi. The allowable stress for $Kl/r = 63.6$ will fall between these two values and, by interpolating, it is found to be 17.08 ksi.

(3) The allowable load on the column is then

$$P = A \times F_a = 19.1 \times 17.08 = 326 \text{ kips}$$

Example 2. A built-up column of A36 steel consists of a W 14 × 320 column core section with two 18 × 1 in. cover plates, as indicated in Fig. 10-4. This is a type of section used in tall buildings where there are extremely heavy loads in the lower-story columns; it has been introduced here to illustrate the determination of the radius of gyration of built-up sections. The pertinent properties of the W 14 × 320 are recorded in the figure. If K, the effective length factor, is 1 and the unbraced height is 20 ft, compute the axial load this built-up column will support.

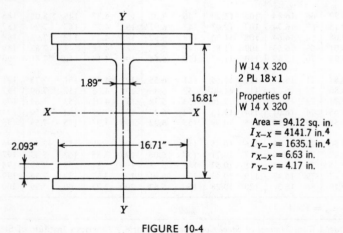

FIGURE 10-4

Solution: The cross-sectional area of the W 14 × 320 is 94.12 sq in., and the area of the two plates is 2 × 18 × 1, or 36 sq in. Consequently, the total cross-sectional area of the built-up section is 94.12 + 36, or 130.12 sq in.

Obviously, the Y–Y axis of the section will have the smaller moment of inertia and consequently the smaller (least) radius of gyration. Therefore, we will first compute r_{Y-Y} for the built-up section. For the W 14 × 320, I_{Y-Y} = 1635.1 in.[4] For an 18 × 1 plate about the axis marked Y–Y, $I = bd^3/12 = (1 \times 18^3)/12 = 486$ in.[4]; and for the two plates, $I_{Y-Y} = 2 \times 486$, or 972 in.[4] Thus, I_{Y-Y} for the entire built-up section is 1635.1 + 972, or 2607.1 in.[4]

Referring to Art. 5-7,

$$r = \sqrt{\frac{I}{A}}$$

then

$$r = \sqrt{\frac{2607.1}{130.12}} \quad \text{and} \quad r = 4.48 \text{ in.}$$

and

$$Kl/r = 1 \times \frac{20 \times 12}{4.48} \quad \text{and} \quad Kl/r \text{ (the slenderness ratio)} = 53.5$$

Table 10-1 shows that the allowable unit stress for a slenderness ratio of 53.5 lies between 17,990 psi and 18,080 psi. By interpolation, F_a, the allowable unit stress, is 18,035 psi.

Then $P = A \times F_a$ and $P = 130.12 \times 18,035 = 2,346,000$ lb, the allowable axial load on the built-up column section.

Note: In the following problems, assume that A36 steel is used and that K, the effective length factor, is 1.0.

Problem 10-6-A.* Compute the allowable axial load on a W 10 × 89 column with an unbraced height of 15 ft.

Problem 10-6-B. A W 12 × 65 used as a column has an unbraced height of 22 ft. Compute the allowable axial load.

Problem 10-6-C.* An angle 5 × 3½ × ⅝ is used as a strut 7 ft long. Compute the allowable axial load.

Problem 10-6-D. A built-up column section is composed of a W 14 × 320 with two 22 × 2 in. cover plates. If the unbraced length is 18 ft, compute the allowable axial load. (For properties of the W 14 × 320, see Fig. 10-4.)

10-7 Design of Steel Columns

In practice, the design of steel columns is accomplished largely by the use of safe load tables; if such tables are not available, design is carried out by the trial method. Data include the load and length of the column. The designer selects a trial cross section on the basis of his experience and judgment and, by means of a column formula, computes the allowable load that it can support. If this load is less than the actual load the column will be required to support, the trial section is too small and another section is tested in a similar manner.

Table 10-2 gives allowable loads on a number of column sections. It has been compiled from more extensive tables in the AISC Manual, and the loads are computed in accordance with the formulas in Art. 10-5. Note particularly that these allowable loads are for main members of A36 steel. Loads to the right of the heavy vertical lines are for Kl/r ratios between 120 and 200. The significance of the bending factors, given at the extreme right of the table, is considered in Art. 10-11.

To illustrate the use of Table 10-2, refer to Example 1 of Art. 10-6. This problem asked for the allowable load that could be supported by a W 12 × 65 of A36 steel with an unbraced height of 16 ft. Referring to Table 10-2, we see at a glance that the allowable axial load is 326 kips, the same value we found by computation.

Although the designer may select the proper column section by merely referring to the safe load tables, it is well to understand the application of the formula by means of which the tables have been computed. To that end, the *design procedure* is outlined below. When the design load and the length have been ascertained, the following steps are taken:

Step (1): Assume a trial section and note from the table of properties both the cross-sectional area and the least radius of gyration.

Step (2): Compute the slenderness ratio Kl/r, l being the unsupported length of the column in inches. For the value of K, see Art. 10-4.

Step (3): Compute F_a, the allowable unit stress, by using a column formula or by use of Table 10-1.

Step (4): Multiply F_a found in step 3 by the area of the column cross section. This gives the allowable load *on the trial column section.*

Step (5): Compare the allowable load found in step 4 with the design load. If the allowable load on the trial section is less than the design load (or if it is so much greater as to make use of the section uneconomical), try another section and test it in the same manner. The reader should note that, except for assuming a trial section, the above operations were carried out in Example 1 of Art. 10-6.

Problem 10-7-A. Using Table 6-2, select a column section to support an axial load of 148 kips if the unbraced height is 12 ft. A36 steel is to be used, and K, the effective length factor, is 1.0.

Problem 10-7-B. Same data as in Problem 10-7-A except that load to be supported = 258 kips and length = 15 ft.

Problem 10-7-C. Some data as in Problem 10-7-A except that load to be carried = 355 kips and length = 20 ft.

10-8 Steel Pipe Columns

Unfilled steel pipe columns are frequently used as main compression members in building construction. In routine work, they are usually designed by the use of safe load tables.

Table 10-3 gives allowable concentric loads[1] for Standard steel pipe columns. The loads are for main members and are computed by formulas 10-5-1 and 10-5-2, (Art. 10-5), for steel with a yield point of 36 ksi. The diameters at the head of the table are *nominal* diameters, and the actual dimensions may vary slightly from the nominal. In computing the allowable loads, K is 1.0. The AISC

---

[1] The reader should be certain that he understands the significance of *axial* and *concentric* as related to column loading. As used in this book, the terms are synonymous and refer to vertical loads applied at the centroid of the cross section (or balanced about it) so that the resultant load acts along the vertical axis of the column, ideally producing a uniform distribution of compressive stress over the cross section. This is in contradistinction to *eccentric* loading, which is considered in Art. 10-11.

TABLE 10-2. Allowable Axial Loads on Columns—

Nominal depth & width	Weight per ft	Area sq in.	Radius of gyration		Effective length in					
			$X–X$	$Y–Y$	6	7	8	9	10	11
4 × 4	13	3.82	1.72	0.99	62	57	51	45	39	32
5 × 5	16	4.70	2.13	1.26	83	79	74	69	64	58
	18.5	5.43	2.16	1.28	97	92	86	81	75	68
6 × 6	15.5	4.56	2.57	1.46	84	81	77	73	69	65
	20	5.88	2.66	1.51	109	105	100	96	91	85
	25	7.35	2.69	1.53	137	132	126	120	114	108
8 × 8	31	9.12	3.47	2.01	178	174	169	164	159	154
	35	10.3	3.50	2.03	201	197	191	186	180	174
	40	11.8	3.53	2.04	231	225	220	213	207	200
	48	14.1	3.61	2.08	276	270	263	256	249	241
	58	17.1	3.65	2.10	336	328	320	312	303	293
	67	19.7	3.71	2.12	387	379	370	360	350	339
10 × 8	33	9.71	4.20	1.94	189	184	179	173	167	161
	39	11.5	4.27	1.98	224	218	213	206	200	193
10 × 10	49	14.4	4.35	2.54	289	284	279	273	268	262
	54	15.9	4.39	2.56	319	314	308	302	296	290
	60	17.7	4.41	2.57	355	350	343	337	330	323
	66	19.4	4.44	2.58	390	383	377	369	362	354
	72	21.2	4.46	2.59	426	419	412	404	396	387
	77	22.7	4.49	2.60	456	449	441	433	424	415
	89	26.2	4.55	2.63	527	519	510	501	491	480
12 × 10	53	15.6	5.23	2.48	312	307	301	295	288	282
	58	17.1	5.28	2.51	343	337	331	324	317	310
12 × 12	65	19.1	5.28	3.02	389	384	378	373	367	361
	72	21.2	5.31	3.04	432	426	420	414	408	401
	79	23.2	5.34	3.05	473	467	460	453	446	439
	85	25.0	5.38	3.07	510	503	496	489	482	474
	92	27.1	5.40	3.08	553	546	538	530	522	514
	99	29.1	5.43	3.09	593	586	578	570	561	552
	106	31.2	5.46	3.11	637	629	620	611	602	593
14 × 10	61	17.9	5.98	2.45	358	351	345	338	330	322
	68	20.0	6.02	2.46	400	393	385	377	369	360
14 × 12	78	22.9	6.09	3.00	466	460	453	447	439	432
	84	24.7	6.13	3.02	503	496	489	482	475	467
14 × 14½	87	25.6	6.15	3.70	528	523	518	512	506	500
	95	27.9	6.17	3.71	576	570	564	558	552	545
	103	30.3	6.21	3.72	625	619	613	606	599	592

Loads to right of heavy lines are for main members with Kl/r ratios between 120 and 200.

* Compiled from data in the 7th Edition of the *Manual of Steel Construction*. Courtesy American Institute of Steel Construction.

Selected W Shapes,* Values in kips for A36 Steel ($F_y = 36$ ksi)

| feet with respect to least radius of gyration | | | | | | | | | | | | Bending factor | |
12	13	14	15	16	17	18	19	20	22	24	26	B_x	B_y
27	23	20	17	15								0.701	2.065
52	46	39	34	30	27	24	21	19				0.551	1.567
62	54	47	41	36	32	28	26	23				0.547	1.534
60	55	50	45	39	35	31	28	25	21	17		0.456	1.412
80	74	68	61	54	48	43	39	35	29	24		0.439	1.328
101	94	86	78	70	62	55	49	45	37	31		0.441	1.308
148	142	136	130	123	117	110	102	95	79	66	57	0.333	0.988
168	162	155	148	141	133	125	117	109	91	76	65	0.332	0.972
193	186	178	170	162	153	144	135	125	105	88	75	0.333	0.976
233	224	215	206	196	186	176	165	154	131	110	94	0.327	0.940
283	273	263	251	240	228	216	203	190	162	136	116	0.329	0.940
328	316	304	292	279	265	251	236	221	190	159	136	0.327	0.921
155	149	142	135	127	120	112	103	95	78	66	56	0.278	1.061
186	178	170	162	154	145	136	126	116	97	81	69	0.273	1.027
255	249	242	235	228	221	213	205	197	180	161	142	0.264	0.775
283	276	268	261	253	245·	236	228	219	200	180	159	0.264	0.769
315	307	299	291	282	273	264	254	244	224	202	178	0.264	0.767
346	337	328	319	310	300	290	279	269	246	222	197	0.264	0.761
378	369	359	349	339	328	317	306	294	270	244	217	0.265	0.760
406	396	385	375	364	352	341	329	316	290	263	233	0.264	0.755
470	458	447	435	422	409	396	382	368	339	308	275	0.263	0.745
275	268	260	252	244	236	227	218	209	189	169	147	0.221	0.813
302	294	286	278	269	260	251	241	231	211	189	165	0.219	0.800
354	348	341	334	326	319	311	303	294	277	259	240	0.218	0.657
394	387	379	371	363	355	346	337	328	309	289	268	0.218	0.655
431	423	415	407	398	389	379	370	360	339	317	294	0.217	0.649
465	457	448	439	430	420	410	400	389	367	344	319	0.216	0.643
505	496	486	476	466	456	445	434	422	399	374	347	0.217	0.643
543	533	523	512	501	490	478	467	454	429	402	374	0.216	0.637
583	572	561	550	539	527	514	502	489	462	433	404	0.216	0.635
314	306	297	288	278	268	258	248	237	214	190	165	0.195	0.833
351	342	332	322	311	301	289	278	266	241	214	186	0.195	0.830
424	416	408	399	390	381	371	362	352	331	309	285	0.190	0.664
458	450	441	431	422	412	402	391	381	358	335	310	0.189	0.659
493	487	480	473	465	458	450	442	434	417	399	381	0.186	0.532
538	531	523	516	508	499	491	482	473	455	436	416	0.185	0.529
585	577	569	560	552	543	534	524	515	495	474	452	0.185	0.527

TABLE 10-3. Typical Column Safe Load Table*

Standard steel pipe
Allowable concentric loads in kips $F_y = 36$ ksi

Nominal dia.	12	10	8	6	5	4	3½	3
Wall thickness	0.375	0.365	0.322	0.280	0.258	0.237	0.226	0.216
Weight per foot	49.56	40.48	28.55	18.97	14.62	10.79	9.11	7.58
6	303	246	171	110	83	59	48	38
7	301	243	168	108	81	57	46	36
8	299	241	166	106	78	54	44	34
9	296	238	163	103	76	52	41	31
10	293	235	161	101	73	49	38	28
11	291	232	158	98	71	46	35	25
12	288	229	155	95	68	43	32	22
13	285	226	152	92	65	40	29	19
14	282	223	149	89	61	36	25	16
15	278	220	145	86	58	33	22	14
16	275	216	142	82	55	29	19	12
17	272	213	138	79	51	26	17	11
18	268	209	135	75	47	23	15	10
19	265	205	131	71	43	21	14	9
20	261	201	127	67	39	19	12	
22	254	193	119	59	32	15	10	
24	246	185	111	51	27	13		
26	238	176	102	43	23			
28	229	167	93	37	20			
30	220	158	83	32	17			
32	211	148	73	29				
34	201	137	65	25				
36	192	127	58	23				
38	181	115	52					
40	171	104	47					

Effective length in feet KL with respect to radius of gyration

Properties

	12	10	8	6	5	4	3½	3
Area A (in.²)	14.6	11.9	8.40	5.58	4.30	3.17	2.68	2.23
I (in.⁴)	279.	161.	72.5	28.1	15.2	7.23	4.79	3.02
r (in.)	4.38	3.67	2.94	2.25	1.88	1.51	1.34	1.16
B (Bending factor)	0.333	0.398	0.500	0.657	0.789	0.987	1.12	1.29
a (Multiply values by 10^6)	41.7	23.9	10.8	4.21	2.26	1.08	0.717	0.447

Heavy line indicates $Kl/r = 120$. Values omitted for $Kl/r > 200$.

* Taken from a more complete set of tables in the 7th Edition of the *Manual of Steel Construction*. Courtesy American Institute of Steel Construction.

Manual contains additional tables giving allowable loads for Extra strong and Double-extra strong steel pipe.

Example. Using Table 10-3, select a steel pipe column to carry a load of 41 kips if the unbraced height is 12 ft. Verify the value given in the table by computing the allowable axial load.

Solution: (1) Entering Table 10-3 with an effective length of 12 ft, we find that a load of 43 kips can be supported by a 4-in. diameter Standard pipe column.

(2) To verify the tabular load, we first note from the Properties listed in the table that the cross-sectional area of this column is 3.17 sq in. and the radius of gyration (the same about any axis through the centroid) is 1.51 in. Then, taking K as 1.0, the slenderness ratio is

$$\frac{Kl}{r} = 1 \times \frac{12 \times 12}{1.51} = 95.3$$

(3) Referring to Table 10-1, we find that F_a (the allowable unit stress) for a slenderness ratio of 95.3 is between 13.6 and 13.48 ksi. By interpolating, $F_a = 13.52$ ksi.

(4) The allowable axial load is

$$P = A \times F_a = 3.17 \times 13.52 = 43 \text{ kips}$$

which agrees with the tabular value.

10-9 Structural Tubing Columns

Steel columns are fabricated from structural tubing in both square and rectangular shapes. Square tubing is available in sizes from 4 to 10 in. and in weights from approximately 12 to 74 lb per lin ft. Rectangular tubing is produced in a variety of cross sections ranging from 5 × 3 in. to 12 × 6 in., with varying wall thicknesses and weights. Many of the shapes are available in two strengths, $F_y = 36$ ksi and $F_y = 46$ ksi. The AISC Manual contains safe load tables for square and rectangular columns of structural tubing based on both strengths. Table 10-4 has been abstracted from these tables to illustrate their general arrangement and use.

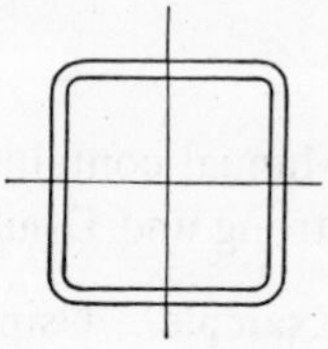

TABLE 10-4.   Typical Column Safe Load Table*

Square structural tubing
Allowable concentric loads in kips

$F_y = 36$ ksi

Nominal size		5 × 5				4 × 4			
Wall thickness		½	⅜	5⁄16	¼	½	⅜	5⁄16	¼
Weight per foot		27.68	21.94	18.77	15.42	20.88	16.84	14.52	12.02
	2	171	135	116	95	127	103	89	74
	3	168	133	114	94	124	100	87	72
	4	164	130	112	92	120	98	84	70
	5	160	127	109	90	116	94	82	68
	6	156	124	107	88	111	91	79	66
	7	151	121	104	86	106	87	76	63
	8	147	117	101	83	101	83	72	60
	9	141	113	98	81	95	79	69	57
	10	136	109	94	78	88	74	65	54
	11	130	105	91	75	82	69	61	51
	12	124	101	87	72	75	64	57	48
	13	118	96	83	69	68	59	52	44
	14	111	91	79	66	60	53	47	40
	15	104	86	75	62	52	47	42	36
	16	97	81	70	59	46	42	37	32
	17	90	75	66	55	41	37	33	29
	18	82	69	61	51	36	33	30	25
	19	74	63	56	47	33	29	27	23
	20	67	57	51	43	29	27	24	21
	22	55	47	42	35	24	22	20	17
	24	46	40	35	30		18	17	14
	26	40	34	30	25				
	28	34	29	26	22				
	30		25	22	19				

$F_y = 36$ ksi — Effective length in feet KL with respect to radius of gyration

Properties

	5 × 5 ½	⅜	5⁄16	¼	4 × 4 ½	⅜	5⁄16	¼
Area A (in.²)	8.14	6.45	5.52	4.54	6.14	4.95	4.27	3.54
I (in.⁴)	25.7	22.0	19.5	16.6	11.4	10.2	9.23	8.00
r (in.)	1.78	1.85	1.88	1.91	1.36	1.44	1.47	1.50
B}Bending factor	0.792	0.733	0.708	0.684	1.08	0.971	0.925	0.885
a}Multiply values by 10⁶	3.84	3.29	2.91	2.47	1.69	1.53	1.37	1.19

Heavy line indicates $Kl/r = 120$. Values omitted for $Kl/r > 200$.

* Taken from a more complete set of tables in the 7th Edition of the *Manual of Steel Construction*. Courtesy American Institute of Steel Construction.

Example. Using the data given for the example on steel pipe columns in Art. 10-8 ($P = 41$ kips and height = 12 ft), select a square structural tubing column from Table 10-4. Compare the two sections.

Solution: Entering Table 10-4 with an effective length of 12 ft, we find that a load of 48 kips can be supported by a square section $4 \times 4 \times \frac{1}{4}$ in. Comparing the two columns:

	Pipe 4 Std.	TS $4 \times 4 \times \frac{1}{4}$
Safe load, kips	43	48
Weight, lb per lin ft	10.79	12.02
Area, sq in.	3.17	3.54
Wall thickness, in.	0.237	0.250
Radius of gyration	1.51	1.50

It will be observed that the circular cross section is somewhat more efficient from the standpoint of use of material since, with a smaller weight per foot than the square column, it has a slightly larger radius of gyration. The choice between round or square columns, however, is normally based on considerations other than structural efficiency.

Problem 10-9-A. A 10-in. steel pipe column of A36 steel has an effective length of 11 ft. Compute its allowable axial load and compare it with the allowable load given in Table 10-3.

Problem 10-9-B. Compute the allowable axial load on a 5-in. steel pipe column of A36 steel whose effective length is 15 ft 6 in.

Problem 10-9-C. A structural tubing column TS $5 \times 5 \times \frac{1}{2}$ of A36 steel is used on an effective length of 12 ft. Compute the allowable concentric load and compare it with the allowable load given in Table 10-4.

Problem 10-9-D.* Using Table 10-4, select the lightest weight square structural tubing column with an effective length of 10 ft 6 in. that will support an axial load of 64 kips.

10-10 Double–Angle Struts

Two angle sections separated by the thickness of a connection plate at each end and fastened together at intervals by fillers and

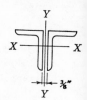

TABLE 10-5. Double-Angle Struts*

Allowable concentric loads in kips. Unequal leg angles

Long legs ⅜ in. back to back of angles

$F_y = 36$ ksi

Size 5 × 3½ — Thickness ¾, ⅝, ½, ⁷/₁₆ — Weight per foot 39.6, 33.6, 27.2, 24.0

Effective length in feet KL with respect to indicated axis

X–X axis KL	¾	⅝	½	⁷/₁₆
0	251	213	173	152
4	231	196	159	140
6	216	184	150	132
8	200	170	139	123
10	181	154	126	112
12	161	137	113	100
14	138	118	97	86
15	126	108	89	79
16	113	97	81	72
18	89	77	64	57
20	72	62	52	46
22	60	51	43	38
24	50	43	36	32
•25	46	40	33	30
26		37	31	27

Y–Y axis KL	¾	⅝	½	⁷/₁₆
0	251	213	173	152
4	230	195	158	139
6	216	183	148	130
8	199	168	136	119
10	180	152	122	107
12	159	133	107	93
14	136	113	90	78
15	123	102	81	70
16	110	91	72	62
18	87	72	57	49
20	70	58	46	39
22	58	48	38	33
24	49	40	32	27
25	45	37		

Size 5 × 3 — Thickness ½, ⁷/₁₆ — Weight per foot 25.6, 22.6

X–X axis KL	½	⁷/₁₆
0	162	143
2	157	138
4	149	132
6	141	124
8	130	115
10	119	105
12	106	94
14	92	82
15	84	75
16	76	68
18	61	54
20	49	44
22	41	36
24	34	31
26	29	26

Y–Y axis KL	½	⁷/₁₆
0	162	143
2	155	137
4	145	128
6	132	117
8	118	104
10	101	89
12	82	72
14	62	54
16	47	41
18	38	33
20	30	26

Size 4 × 3½ — Thickness ⅝, ½, ⁷/₁₆, ⅜, ⁵/₁₆ — Weight per foot 29.4, 23.8, 21.2, 18.2, 15.

X–X axis KL	⅝	½	⁷/₁₆	⅜	⁵/₁₆
0	186	151	133	115	97
2	177	144	128	110	93
4	165	135	119	103	87
6	151	123	109	94	79
8	133	109	97	84	71
10	113	93	83	72	61
12	91	75	67	59	50
14	68	56	50	44	38
16	52	43	38	34	29
18	41	34	30	27	23
20	33	27	25	22	18
21					17

Y–Y axis KL	⅝	½	⁷/₁₆	⅜	⁵/₁₆
0	186	151	133	115	97
2	179	146	129	111	94
4	171	139	123	106	89
6	161	131	116	100	84
8	150	121	107	92	77
10	137	111	97	84	70
12	122	98	87	74	62
14	106	85	75	64	53
15	97	78	68	58	49
16	88	70	61	53	44
18	70	56	49	42	35
20	57	45	39	34	28
22	47	37	33	28	23
24	40	31	27	23	19
25	36	29	25	22	18
26	34	27	23	20	

Properties of 2 angles—⅜ in. back-to-back

	5 × 3½				5 × 3		4 × 3½				
Area A (in.²)	11.6	9.84	8.00	7.05	7.50	6.62	8.59	7.00	6.18	5.34	4.49
r_x (in.)	1.55	1.56	1.58	1.59	1.59	1.60	1.22	1.23	1.24	1.25	1.26
r_y (in.)	1.53	1.51	1.49	1.47	1.25	1.24	1.60	1.58	1.57	1.56	1.55

Heavy line indicates $Kl/r = 120$. Values omitted for $Kl/r > 200$.

* Compiled from data in the 7th Edition of the *Manual of Steel Construction*. Courtesy, American Institute of Steel Construction.

TABLE 10-5. Double-Angle Struts (Continued)

Allowable concentric loads in kips. Unequal leg angles

Long legs ⅜ in. back to back of angles

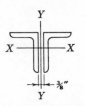

$F_y = 36$ ksi — Effective length in feet KL with respect to indicated axis

Size	4 × 3					3½ × 2½				2½ × 2		
Thickness	⅝	½	⁷⁄₁₆	⅜	⁵⁄₁₆	½	⁷⁄₁₆	⅜	⁵⁄₁₆	⅜	⁵⁄₁₆	¼
Weight per foot	27.2	22.2	19.6	17.0	14.4	18.8	16.6	14.4	12.2	10.6	9.0	7.2

X-X axis

4 × 3:

KL	⅝	½	⁷⁄₁₆	⅜	⁵⁄₁₆
0	172	140	124	107	90
2	164	134	119	103	86
4	154	126	111	96	81
6	140	115	101	88	74
8	124	102	90	78	66
10	106	88	77	67	57
12	85	71	63	55	47
14	64	54	47	42	36
16	49	41	36	32	27
18	39	33	29	25	22
20	31	26	23	20	17
21				19	16

3½ × 2½:

KL	½	⁷⁄₁₆	⅜	⁵⁄₁₆
0	119	105	91	77
2	113	100	86	73
4	104	92	80	67
6	93	82	71	60
8	79	70	61	52
10	64	57	50	42
11	56	49	43	37
12	47	42	37	31
14	35	31	27	23
16	26	23	21	18
17	23	21	18	16
18	21	19	16	14

2½ × 2:

KL	⅜	⁵⁄₁₆	¼
0	67	57	46
2	61	52	42
3	58	49	40
4	53	45	37
5	48	41	34
6	42	36	30
7	36	31	26
8	30	26	21
9	23	20	17
10	19	16	14
11	16	14	11
12	13	11	9
13			8

Y-Y axis

4 × 3:

KL	⅝	½	⁷⁄₁₆	⅜	⁵⁄₁₆
0	172	140	124	107	90
2	165	135	119	103	86
4	156	127	112	97	81
6	144	117	103	89	75
8	130	105	93	80	67
10	115	92	81	70	58
12	97	77	68	58	48
13	88	69	61	52	43
14	78	61	53	45	37
16	60	47	41	35	29
18	47	37	32	27	23
20	38	30	26	22	18
21	35	27	24	20	17
22	32	25	21		

3½ × 2½:

KL	½	⁷⁄₁₆	⅜	⁵⁄₁₆
0	119	105	91	77
2	113	100	87	73
4	105	92	80	67
6	94	83	72	60
8	82	72	62	52
10	68	59	50	42
11	60	52	44	37
12	51	44	37	31
14	38	32	28	23
16	29	25	21	17
18	23	20	17	14

2½ × 2:

KL	⅜	⁵⁄₁₆	¼
0	67	57	46
2	63	53	43
3	60	51	41
4	57	48	39
5	53	45	36
6	49	41	33
7	45	38	30
8	40	34	27
9	35	29	23
10	30	24	19
11	24	20	16
12	21	17	13
13	18	14	11
14	15	12	10
15	13	11	9
16	12		

Properties of 2 angles—⅜ in. back-to-back

	4 × 3					3½ × 2½				2½ × 2		
Area A (in.²)	7.97	6.50	5.74	4.97	4.18	5.50	4.87	4.22	3.55	3.09	2.62	2.13
r_x (in.)	1.23	1.25	1.25	1.26	1.27	1.09	1.09	1.10	1.11	0.768	0.776	0.784
r_y (in.)	1.36	1.33	1.32	1.31	1.30	1.14	1.12	1.11	1.10	0.961	0.948	0.935

Heavy line indicates $Kl/r = 120$. Values omitted for $Kl/r > 200$.

welds or rivets, are commonly used as compression members in roof trusses. These members, whether or not in a vertical position, are called struts; their size is determined in accordance with the requirements and formulas for columns given in Art. 10-5. To assure that the angles act as a unit, the intermittent connections are made at intervals such that the slenderness ratio l/r of either angle between fasteners does not exceed the governing slenderness ratio of the built-up member. The least radius of gyration r is used in computing the slenderness ratio of each angle.

The AISC Manual contains safe load tables for struts of two angles with $\frac{3}{8}$-in. separation back-to-back. Three series are given: equal-leg angles, unequal-leg angles with short legs back-to-back, and unequal-leg angles with long legs back-to-back. Table 10-5 has been abstracted from the latter series and lists allowable loads with respect to both the X–X and Y–Y axes. The smaller (least) radius of gyration gives the smaller allowable load and, unless the member is braced with respect to the weaker axis, this is the tabular load to be used. The usual practice is to assume K equal to 1.0. The following example shows how the loads in the table are computed.

Example. Two $5 \times 3\frac{1}{2} \times \frac{1}{2}$ angle sections spaced with their long legs $\frac{3}{8}$ in. back-to-back are used as a compression member. If the member is A36 steel and has an effective length of 10 ft, compute the allowable axial load.

Solution: At the bottom of Table 10-5 under Properties, we find that the area of the two-angle member is 8.0 sq in. and that the radii of gyration are $r_x = 1.58$ in. and $r_y = 1.49$ in. Using the smaller r, the slenderness ratio is

$$\frac{Kl}{r} = 1 \times \frac{10 \times 12}{1.49} = 80.5$$

Referring to Table 10-1 and interpolating, we find that $F_a = 15.3$ ksi for a slenderness ratio of 80.5, making the allowable load

$$P = A \times F_a = 8.0 \times 15.3 = 122 \text{ kips}$$

This value is, of course, readily verified by entering Table 10-5 under "Y–Y Axis" with an effective length of 10 ft and then proceeding horizontally to the column giving loads for the $5 \times 3\frac{1}{2} \times \frac{1}{2}$ angle.

The design of double-angle struts as compression members for roof trusses is considered in Chapter 13.

Problem 10-10-A.* A double-angle strut 8 ft long is composed of two angles 4 × 3 × ⅜ in. with the long legs ⅜ in. back-to-back. If the member is fabricated from A36 steel, compute the allowable concentric load.

Problem 10-10-B. Using Table 10-5, select a double-angle strut that will support a concentric load of 50 kips if the effective length is 10 ft.

10-11 Eccentrically Loaded Columns

The columns previously discussed have been axially or concentrically loaded.[2] It frequently happens, however, that in addition to the axial load the column may be subjected to bending stresses that

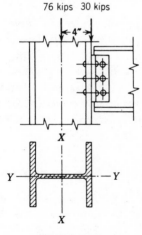

76 kips 30 kips

FIGURE 10-5

result from eccentric loads. Figure 10-5 indicates a column having both a concentric and an eccentric load. The design of eccentrically loaded columns is accomplished by testing trial sections. As an aid to design, it will be found convenient to convert the axial and eccentric loads into a single equivalent axial load. Having done this, the safe load tables may be used to select the trial section.

[2] See footnote to Art. 10-8.

10-12 Bending Factors

On the right-hand side of Table 10-2, the *bending factors* B_x and B_y are listed. The bending factor is the area of the cross section divided by its section modulus. Because there are two major section moduli, there are two bending factors, B_x for the X–X axis and B_y for the Y–Y axis. For example, the area of a W 10 × 49 is 14.4 sq in., and the section moduli with respect to the X–X and Y–Y axes are 54.6 in.^{3} and 18.6 in.^{3}, respectively. Then

$$B_x = \frac{A}{S_x} = \frac{14.4}{54.6} = 0.264$$

and

$$B_y = \frac{A}{S_y} = \frac{14.4}{18.6} = 0.775$$

Note that these are the values given in Table 10-2.

Bending factors are used to convert the effect of eccentricity to an equivalent axial load. To accomplish this, the *bending moment* resulting from the eccentric load is multiplied by the appropriate bending factor. Then, *the total equivalent axial load* (P') *is equal to the sum of the axial load and the eccentric load plus the product of the bending moment due to the eccentric load and the appropriate bending factor.*

10-13 Trial Section for Eccentrically Loaded Columns

The trial section used in designing a column subjected to both axial and eccentric loads may be established by first finding an *approximate* equivalent axial load. This procedure is illustrated in the following example.

Example. An 8-in. column of A36 steel with an unsupported height of 13 ft supports an axial load of 76 kips and a load of 30 kips applied 4 in. from the X–X axis. The arrangement is shown in Fig. 10-5. Determine the trial column section.

Solution: (1) The bending moment produced by the eccentric load is

$$M = 30 \times 4 = 120 \text{ kip-in.}$$

However, only the general dimensions of the section are known at this point, so we do not know the exact value of the bending factor.

(2) Referring to the 8-in. column sections in Table 10-2, tentatively select a bending factor of 0.333. This may be revised later. Then the bending moment multiplied by the bending factor is 120 × 0.333 = 39.9 or 40 kips, which is an equivalent axial load for the eccentric load.

(3) Now in accordance with the principle stated in Art. 10-12, the approximate total equivalent load on the column is

$$P' = 76 + 30 + 40 = 146 \text{ kips}$$

Referring to Table 10-2 again, we find that a W 8 × 35 will carry 162 kips with an effective length of 13 ft, and a W 8 × 31 is listed for 142 kips.

Since the foregoing procedure gives results that are approximate on the safe side, the W 8 × 35 could be the accepted section. However, if it is desired to determine the lightest weight section that can be used, the W 8 × 31 should be investigated for compliance with the AISC Specification requirements for design of columns with combined loading. This is not a simple procedure, and the diversity of factors involved makes it inadvisable to include treatment of these complex requirements in a book of this scope. Reference to the AISC Manual is recommended for readers who desire to study the complete specification requirements covering combined axial compression and bending in columns.

The selection of a trial section by the equivalent axial load method is always conservative—increasingly so as the ratio of the eccentric load to the axial load and the column slenderness ratio increase. Nevertheless, for many situations that arise in routine practice, the trial section determined by this method may be taken as the accepted section.

In conventional building construction, it is assumed that the effect of an eccentric load disappears at each story height. Consequently, in the above example, the design load that the W 8 × 35 transmits to the column in the story below or to a base plate is not 146 kips but 106 kips plus the column weight. In this example, only one eccentric load was given. If, in addition, there is an eccentric load about the $Y-Y$ axis, its equivalent axial load plus the magnitude of the load is added to the 146 kips to determine the total approximate equivalent axial load on the column.

10-14 Column Splices

In the construction of steel frames of buildings, it is common practice to use columns of two-story lengths. This results in a greater cross-sectional area in the upper story than the load requires, but the cost of the excess material is offset by the saving in fabricating costs for the extra splice. When columns of two-story lengths are employed, the load on the lower story length is used in determining the required cross section of the column.

In order not to conflict with the beam and girder connections, column splices are made 2 ft or more above the floor level. In general, splices are made by means of plates $\frac{3}{8}$ or $\frac{1}{2}$ in. thick that are bolted, welded, or riveted to the flanges of the columns, as indicated in Fig. 10-6 and Fig. 12-9. The splice plates are not designed to resist compressive stresses; their function is to hold the column sections in position. Since the upper column transmits its load directly to the column below, the surfaces in contact should be milled to provide full bearing areas. When the upper column has a smaller width than the supporting column, filler plates are used as indicated in Fig. 10-6b. If the difference in width is so great that a full bearing area

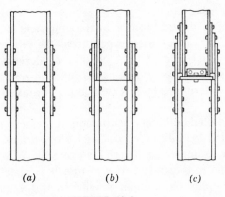

(a)   (b)   (c)

FIGURE 10-6

between the columns is not achieved, a horizontal plate is used as in Fig. 10-6c.

10-15 Column Base Plates

Steel columns generally transfer their loads to the foundation bed by means of reinforced concrete footings. As the allowable compressive strength of concrete is considerably less than the actual unit stresses in the steel column, it is necessary to provide a steel base plate under the column to spread the load sufficiently to prevent over-stressing of the concrete. The typical arrangement is shown in Fig. 10-7.

Rolled steel bearing plates used for column bases may be obtained in a great variety of sizes. The lengths and widths have dimensions usually in even inches and, for the sizes commonly used in building construction, the plates may be obtained in ⅛-in. increments of thickness. So that the column load may be distributed uniformly over the base plate, it is important that the column and plate be in absolute contact. Rolled steel bearing plates more than 2 in. thick but not more than 4 in. thick may be straightened by pressing or planing. Material more than 4 in. in thickness must be planed on the upper surface. The under surface need not be planed since a full

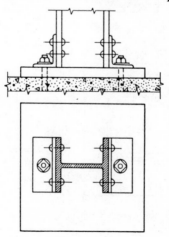

FIGURE 10-7

bearing contact is obtained by using a layer of cement grout on the concrete surface.

Steel columns are usually secured to the foundation by steel bolts embedded in the concrete that pass through the base plate and angles riveted or welded to the column flange. See Figs. 10-7 and 12-9l and m. For light columns, the angles are often omitted, and the base plate is secured to the column by fillet welds.

The first step in the design of a base plate is to determine its area. This is accomplished by use of the basic formula

$$A = \frac{P}{F_p}$$

in which P = the column load in pounds

F_p = the allowable bearing value of the concrete in psi. The AISC specification gives this stress as $0.25f'_c$ when the entire area of the concrete support is covered and $0.375f'_c$ when only one third of the area is covered. A concrete commonly used has 3000 psi for the value of f'_c; hence, F_p = either 750 psi or 1125 psi.

and $A = B \times N$ = the area of the base plate in sq in. see Fig. 10-8a.

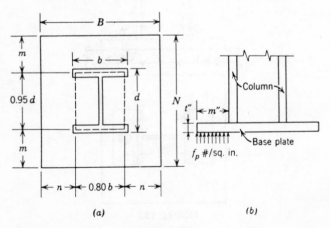

(a)   (b)

FIGURE 10-8

The column load is assumed to be uniformly distributed over the dotted rectangle shown in Fig. 10-8a. In addition, the base plate is assumed to exert a uniform pressure on the concrete foundation.

After the minimum required area of the base plate has been found, B and N are established so that the projections m and n are approximately equal.

The final step is to determine m and n in inches and to use the larger value in computing the thickness of the plate by use of the formula

$$t = \sqrt{\frac{3f_p m^2}{F_b}} \quad \text{or} \quad t = \sqrt{\frac{3f_p n^2}{F_b}}$$

in which t = the thickness of the bearing plate, in inches

f_p = the *actual* bearing pressure on the foundation, in psi

and F_b = the allowable bending stress in the base plate, in psi.

The AISC specification gives the value of F_b as $0.75F_y$, F_y being the yield point of the steel plate. Thus, for A36 steel, $F_y = 36,000$ psi and $F_b = 0.75 \times 36,000$, or $F_b = 27,000$ psi.

The foregoing formulas are readily derived. Refer to Fig. 10-8b and consider a strip of base plate 1 in. wide and t in. thick projecting m in. beyond the face of the column flange. Since the upward pressure on the steel strip is f_p pounds per square inch, the strip is actually a cantilever bending upward. Hence, the bending moment at the face of the column is

$$M = f_p m \times \frac{m}{2} = \frac{f_p m^2}{2}$$

For this rectangular strip

$$\frac{I}{c} = \frac{bd^2}{6} \qquad \text{(Art. 5-5)}$$

and since $b = 1$ in. and $d = t$ in.,

$$\frac{I}{c} = \frac{1 \times t^2}{6} = \frac{t^2}{6}$$

As the strip is in flexure

$$\frac{M}{f} = \frac{I}{c} \qquad \text{(Art. 6-1)}$$

and if $f = F_b$,

$$\frac{f_p m^2}{2} \times \frac{1}{F_b} = \frac{t^2}{6} \quad \text{and} \quad t^2 = \frac{6 f_p m^2}{2 F_b} \quad \text{or} \quad t = \sqrt{\frac{3 f_p m^2}{F_b}}$$

If A36 steel is used for the plate, $F_b = 27{,}000$ psi (Table 3-2), and the two thickness formulas become

$$t = \sqrt{\frac{f_p m^2}{9000}} \quad \text{and} \quad t = \sqrt{\frac{f_p n^2}{9000}}$$

Example. A W 14 × 74 is used as a column to support an axial load of 435,000 lb. If the allowable bearing strength of the concrete foundation is $F_p = 750$ psi, design a base plate of A36 steel.
Solution: (1) The required area of the plate is

$$A = P/F_p = 435{,}000/750 = 580 \text{ sq in.}$$

If we establish B as 22 in., N becomes $580/22 = 26.3$, say 27 in.

(2) Referring to Table 1-1, we find for a W 14 × 74 that the depth is 14.19 in. and the flange width is 10.072 in. Then, using Fig. 10-8, the values of m and n are determined as follows:

$$0.8 \times 10.072 = 8.05 \text{ in.} \qquad n = \frac{22 - 8.05}{2} \quad \text{and} \quad n = 6.97 \text{ in.}$$

$$0.95 \times 14.19 = 13.5 \text{ in.} \qquad m = \frac{27 - 13.5}{2} \quad \text{and} \quad m = 6.75 \text{ in.}$$

(3) The actual unit pressure on the under side of the plate is

$$f_p = \frac{435{,}000}{22 \times 27} = 732 \text{ psi}$$

(4) Because n (6.97 in.) is greater than m (6.75 in.), it is used in the formula for thickness. Thus,

$$t = \sqrt{\frac{f_p n^2}{9000}} = \sqrt{\frac{732 \times 6.97 \times 6.97}{9000}} = 1.99 \text{ in.}$$

say 2 in. Consequently, we accept a column base plate 22 × 27 × 2 in.

Problem 10-15-A.* Design a base plate of A36 steel for a W 12 × 40 column that supports an axial load of 229 kips. The footing on which the base plate rests is concrete, for which $F_p = 750$ psi.

Problem 10-15-B. Design a base plate for the W 12 × 40 column given in Problem 10-15-A, assuming that $F_p = 1125$ psi instead of 750 psi.

10-16   Grillage Foundations

Before reinforced concrete was used universally as a building material, it was customary to employ steel grillage for sizable foundations. The purpose of a grillage is to distribute column loads over the required area of the foundation bed. Because reinforced concrete footings are more economical, they have, except in certain situations, supplanted steel grillages for use as column footings. For

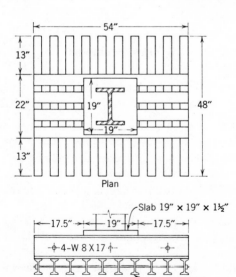

FIGURE 10-9

very heavy column loads, grillage beams may be lighter and more economical than rolled steel base plates on concrete.

Grillage foundations consist of one or more tiers of steel beams arranged as shown in Fig. 10-9, which represents a typical design for a column load of 216 kips. Ample space is provided between the flanges for placing and rodding the concrete. The concrete fills all the spaces between beams, and the entire system of beams is enclosed in at least 4 inches of concrete. The concrete helps the beams to distribute the load to the foundation bed and serves also to protect the steel from corrosion. Pipe separators hold the beams in position, or welded diaphragms are used as separators.

The general structural action of a grillage foundation may be discerned from Fig. 10-9. By means of the base plate, the column load is transferred to the first tier of beams, which then spreads the load over the lower tier at a reduced unit bearing pressure. The bearing pressure on the bottom of the grillage is limited to the bearing capacity of the concrete, inasmuch as a thin concrete mat is usually placed under the grillage even where the supporting material is bedrock.

Review Problems

Note: In the following problems, assume that the columns and base plates are A36 steel and that K, the effective length factor, is 1.0.

Problem 10-16-A. A W 10 × 60 is used as a main compression member. If its effective length is 16 ft, what safe load will this column support?

Problem 10-16-B. A W 6 × 25 is used as a column with an effective length of 12 ft 6 in. Compute its allowable concentric load.

Problem 10-16-C. The effective length of a W 8 × 40 used as a column is 22 ft. Compute its allowable axial load.

Problem 10-16-D.* What is the lightest weight wide flange section that can be used to support an axial compressive load of 250 kips if the effective length is 18 ft?

Problem 10-16-E.* An S 10 × 25.4 is used as a column with an effective length of 8 ft. Compute its allowable axial load.

Problem 10-16-F. A W 14 × 320 has 20 × 1½ plates welded to its flanges. Determine the allowable axial load when used as a column with an effective length of 16 ft. (The properties of the W 14 × 320 are listed in Fig. 10-4.)

Problem 10-16-G.   Two 12 × ½ plates are riveted to the flanges of an S 12 × 31.8 to constitute a built-up member for use as a column. If its effective length is 13 ft, compute the allowable concentric load.

Problem 10-16-H.*   A single angle 8 × 8 × 1 is used as a strut with an effective length of 7 ft. Compute the allowable axial load.

Problem 10-16-I.   An 8-in. column of A36 steel has an axial load of 90 kips and a load of 30 kips applied 4 in. from the $X–X$ axis. The arrangement is similar to that shown in Fig. 10-5. The unsupported length of the column is 13 ft, and the column end conditions are as shown in Fig. 10-3d. Is a W 8 × 35 acceptable for this loading?

Problem 10-16-J.*   Design a column base plate for a W 8 × 31 column that is supported on a concrete foundation for which the allowable bearing capacity is 750 psi. The load on the column is 178 kips.

Problem 10-16-K.   Design a column base for the W 8 × 31 of Problem 10-16-J, assuming the allowable bearing value on the concrete foundation is 1125 psi instead of 750 psi.

Bolted and Riveted Connections

IIIIIIIIIIIIIIIIIIIIIIIIIIIIIIIIIIIIIIIIIIIIIII

11-1 General

Most connections in modern steel building construction are made with bolts or welds. Riveting, once the dominant method of joining structural steel members, is seldom used today. In our discussion of bolts and rivets as fasteners, however, we will consider the principles underlying riveted construction first because much of the technology related to bolted connections has developed from the long experience with rivets.

11-2 Riveting

The simplest type of riveted joint is illustrated in Fig. 11-1a and b. The two bars to be connected are first punched or drilled and are held securely one on the other so that the holes are in alignment. A heated rivet is placed in the hole and a bucking-up tool is pressed against the rivet head. The projecting shank is then covered by a power riveter that delivers rapid blows, filling the hole, deforming the shank, and forming the head. On cooling, there is a slight shrinkage in the length of the rivet and the two plates are thus drawn tightly together.

The design of a riveted joint in many cases is a comparatively simple matter, and the computations are not involved. The computations generally, however, are based on certain assumptions that are not in strict accordance with reality. These assumptions are as follows:

1. The friction between the joined plates is ignored.
2. The stress transferred from one plate to another is equally distributed to all rivets in the joint.
3. The tensile stress in the net sections of the plates is uniform for each square inch of cross section.
4. The bending stresses in the rivets are ignored.
5. The rivets fill the holes completely, and there is no slip between the plates.

The size of the rivet used is determined by the class of work, the thicknesses of the materials to be connected, and the stress to be transmitted across the joint. For structural steelwork in buildings, ¾ and ⅞ in. diameter rivets are the sizes most commonly used. If possible, more than one size rivet on a structure should be avoided.

The heads of rivets are generally *button heads*. When adjacent members are so close that button heads provide insufficient clearance,

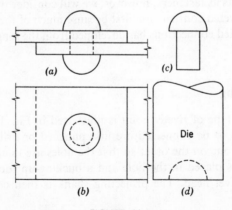

FIGURE 11-1

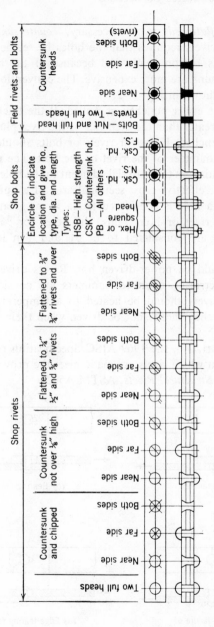

FIGURE 11-2 Conventional signs for rivets and bolts. Taken from the 2nd Edition of *Structural Steel Detailing*. Courtesy American Institute of Steel Construction.

the heads are *flattened* or, if necessary, *countersunk*. The use of countersunk rivets requires that the holes be reamed. They are not so strong as button heads and because of the added labor required for reaming are more expensive. The conventional symbols for riveting are shown in Fig. 11-2.

Rivet holes are either punched or drilled. If the thickness of the material is not greater than the diameter of the rivet plus $\frac{1}{8}$ in., the holes may be punched. Holes for rivets or bolts should be $\frac{1}{16}$ in. larger than the diameter of the rivet or the bolt. The punching of structural steel damages the material adjacent to the hole. Therefore, in computing the effective cross-sectional area of a punched plate, the diameter of a hole is considered to be $\frac{1}{8}$ in. greater than the diameter of the rivet or bolt. Thus, the holes for $\frac{3}{4}$ and $\frac{7}{8}$ in. rivets and bolts are considered to be $\frac{7}{8}$ and 1 in in diameter, respectively.

All rivets should be power-driven hot. Rivets driven by pneumatically or electrically operated hammers are considered to be power-driven. Rivets should be heated to a temperature not exceeding 1950°F; they should not be driven when their temperature is below 1000°F.

As noted in Art. 3-6, the 1969 AISC Specification requires that the material from which rivets are made meet the provisions of the *Specification for Structural Rivets*, ASTM A502.

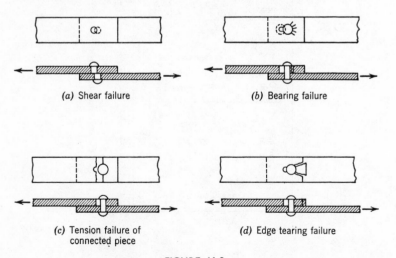

(a) Shear failure   (b) Bearing failure

(c) Tension failure of   (d) Edge tearing failure
     connected piece

FIGURE 11-3

11-3 Failure of Riveted Joints

A riveted joint may fail in one of several ways, as illustrated in Fig. 11-3:

(a) By shearing of the rivet
(b) By crushing of the rivet or the metal on which it bears
(c) By tension in the net sections of the connected members
(d) By tearing at the edge

Failures by tension in the net sections and by tearing out at the edges are prevented by providing ample material between the rivet holes and a sufficient edge distance. If the minimum pitches and edge distances (Arts. 11-11 and 11-12) are maintained, such failures will be avoided.

Shearing and bearing failures, the result of excessive stresses on the rivets, are avoided by providing a sufficient number of rivets to keep the stresses within the allowable limits. As noted earlier, the friction between the joined plates is neglected; this is done because the degree of clamping effect from rivet shrinkage during cooling cannot be measured with precision. The edges of the plates are in

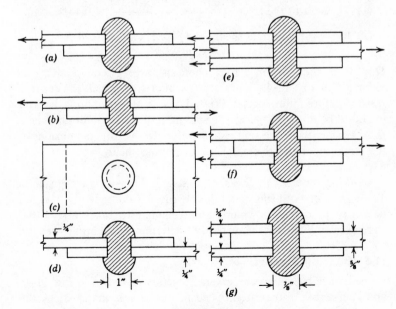

FIGURE 11-4

bearing against the rivet, and the shank of the rivet resists shear. This is called a *bearing-type connection* in contrast to the *friction-type connection* attained with high-strength bolts (Art. 11-9).

11-4 Shear Strengths of Rivets

Consider two bars riveted together as shown in Fig. 11-4a and c. If tensile or compressive forces are applied to the bars, there is a tendency for the bars to slide one on the other and for the rivet to fail by shearing at the plane of contact between the two bars. This failure is indicated in Fig. 11-4b. The rivet is said to be in *single shear*, since it may fail on a single plane. If F_v is the allowable unit shearing stress on the rivet cross section in kips per square inch and D is the diameter of the rivet in inches, the cross-sectional area of the rivet is $\pi D^2/4$, and the strength (allowable value) of the rivet in *single shear* is $(\pi D^2/4) \times F_v$. It is the *nominal* diameter of the rivet that is used in such computations.

Example. Compute the allowable value of a ¾-in. diameter rivet in single shear if the allowable shearing stress is 15 ksi.
Solution:

(1) The area of the rivet $= \dfrac{\pi D^2}{4} = \dfrac{3.14 \times 0.75^2}{4} = 0.4418$ sq in.

(2) Rivet value (single shear) $= F_v \times A = 15 \times 0.4418$
$$= 6.627 \text{ kips}$$

Double Shear. When a rivet connects three pieces of metal, as indicated in Fig. 11-4e, there are *two* planes on which the shear is resisted before failure occurs (Fig. 11-4f). In this situation, the rivet is said to be in double shear. The strength of a rivet in double shear is, of course, twice the single shear value. In the foregoing example, the strength of the rivet, if used in double shear, would be 2 × 6.63 = 13.25 kips.

Working values for rivets of different diameters are given in Table 11-2 for both single and double shear. Consequently, it is not necessary to carry out computations similar to those above that were used to illustrate the method of determining shearing strength.

11-5 Bearing Strength of Rivets

Consider again the two bars riveted together as shown in Fig. 11-4a and c. If tensile forces are applied to the bars as indicated by the arrows, pressure is exerted by the bars on the shank of the rivet. The stresses produced by this pressure are *bearing stresses*. Because the surfaces in contact are curved (cylindrical), the stresses transferred from the plates to the rivets are not distributed uniformly. However, for purposes of design, these stresses are assumed to be uniformly distributed over the *projected* area of contact between the rivet shank and the plate—that is, the area of a rectangle with sides equal to the nominal diameter of the rivet and the thickness of the plate.

When two pieces of metal are riveted together as shown in Fig. 11-4a and c, the rivet is in single shear and in single-shear bearing. If the joint is arranged as in Fig. 11-4e, the rivet is in double shear and in double-shear bearing (enclosed bearing). The allowable unit bearing stress is the same for either condition.

Example 1. Compute the bearing capacity of a $\frac{3}{4}$-in. rivet used to connect the two $\frac{1}{4}$-in. plates shown in Fig. 11-4d if the allowable unit bearing stress $F_p = 48.6$ ksi.
Solution:
   (1) Bearing area = $0.75 \times 0.25 = 0.1875$ sq in.
   (2) Rivet value ($\frac{1}{4}$-in. bearing) = $F_p \times A = 48.6 \times 0.1875$
                                              = 9.11 kips

Working values for rivets, as determined by their bearing capacity in different thicknesses of metal, are tabulated in Table 11-2 in order to avoid the necessity of making computations similar to those in the foregoing example.

Example 2. Compute the bearing capacity of a $\frac{7}{8}$-in. rivet in the connection shown in Fig. 11-4g if the allowable unit bearing stress is 48.6 ksi. Compare this value with the rivet's shear capacity and state the *controlling value* in this joint.
Solution: (1) To determine the bearing area, we note that the enclosed plate acting to the right is $\frac{5}{8}$ in. thick and that the combined thickness of the two plates acting to the left is $\frac{1}{4} + \frac{1}{4} = \frac{1}{2}$ in. Therefore, the controlling bearing area is equal to the nominal diameter of the rivet multiplied by the thinner of these two thicknesses, or
                 Bearing area = $0.875 \times 0.5 = 0.4375$ sq in.
   (2) Rivet value ($\frac{1}{2}$-in. bearing) = $F_p \times A = 48.6 \times 0.4375$
                                              = 21.3 kips

TABLE 11-1. Allowable Tension and Shear Unit Stresses for Rivets and Bolts*
(kips per square inch)

Description of fastener	Tension (F_t)	Shear (F_v) Friction-type connections	Shear (F_v) Bearing-type connections
A502, Grade 1, hot-driven rivets	20.0		15.0
A502, Grade 2, hot-driven rivets	27.0		20.0
A307 bolts	20.0[1]		10.0
Threaded parts[3] of steel meeting the requirements of Sect. 1.4.1	$0.60F_y$[1]		$0.30F_y$
A325 and A449 bolts, when threading is *not* excluded from shear planes	40.0[2]	15.0	15.0
A325 and A449 bolts, when threading is excluded from shear planes	40.0[2]	15.0	22.0
A490 bolts, when threading is *not* excluded from shear planes	54.0[2,4]	20.0	22.5
A490 bolts, when threading is excluded from shear planes	54.0[2,4]	20.0	32.0

* Reproduced from the 7th Edition of the *Manual of Steel Construction*. Courtesy American Institute of Steel Construction.

[1] Applied to tensile stress area equal to $0.7854\left(D - \dfrac{0.9743}{n}\right)^2$ where D is the major thread diameter and n is the number of threads per inch.

[2] Applied to the nominal bolt area.

[3] Since the nominal area of an upset rod is less than the stress area, the former area will govern.

[4] Static loading only.

This value may be found directly from Table 11-2 by entering the column for $\frac{7}{8}$-in. rivets and moving down page to the line for $\frac{1}{2}$-in. bearing thickness.

(3) The shear capacity of this rivet (which is in double shear) is given in Table 11-2 as 18.04 kips. Since this value is less than the bearing capacity of 21.3 kips, the *controlling value* of the rivet in this joint is 18.04 kips.

TABLE 11-2. Allowable Loads for Rivets and Threaded Fasteners in Bearing-Type Connections in kips*

Unit Shearing Stress = 15.0 ksi

Unit Bearing Stress = 48.6 ksi

Diameter		$\frac{5}{8}$ in.	$\frac{3}{4}$ in.	$\frac{7}{8}$ in.	1 in.	$1\frac{1}{8}$ in.
Area, sq in.		0.0368	0.4418	0.6013	0.7854	0.9940
Single shear		4.60	6.63	9.02	11.78	14.19
Double shear		9.20	13.25	18.04	23.56	29.82
Bearing						
Thickness of plate in inches	$\frac{1}{4}$	7.59	9.11	10.6	12.2	13.7
	$\frac{5}{16}$	9.49	11.4	13.3	15.2	17.1
	$\frac{3}{8}$	11.4	13.7	15.9	18.2	20.5
	$\frac{7}{16}$	13.3	15.9	18.6	21.3	23.9
	$\frac{1}{2}$	15.2	18.2	21.3	24.3	27.3
	$\frac{9}{16}$	17.1	20.5	23.9	27.3	30.8
	$\frac{5}{8}$	19.0	22.8	26.6	30.4	34.2

* Compiled from data in the 7th Edition of the *Manual of Steel Construction*. Courtesy American Institute of Steel Construction.

11-6 Allowable Stresses for Rivets

The allowable unit stresses of 15 ksi for shear and 48.6 ksi for bearing, used in the examples of Arts. 11-4 and 11-5, are those prescribed by the 1969 AISC Specification for ASTM A502, Grade 1, hot-driven rivets in bearing-type connections. Table 11-1 is a summary of allowable unit stresses in shear and tension for rivets and bolts. The allowable bearing stress, F_p, on the projected area of rivets and bolts in bearing-type connections is

$$F_p = 1.35 \times F_y$$

where F_y is the yield stress of the connected part. For A36 steel,

$$F_p = 1.35 \times 36 = 48.6 \text{ ksi}$$

and, as noted earlier, this value applies to both single-shear bearing and enclosed bearing. The rivet capacities listed in Table 11-2 have been computed using a shearing stress of 15 ksi and a bearing stress of 48.6 ksi.

Rivets are not often used in positions where direct tensile stresses are developed, except in some types of brackets supporting eccentric loads on columns. When they are used in tension, the stresses given in Table 11-1 control.

Problem 11-6-A. Two $\frac{1}{4}$-in. plates are riveted together with 1-in. diameter rivets (Fig. 11-4d). Determine whether the allowable value of one rivet is controlled by shear or bearing, and state the controlling value.

Problem 11-6-B. In a joint similar to that shown in Fig. 11-4d, two $\frac{5}{16}$-in. plates are fastened together with $\frac{7}{8}$-in. rivets. Determine the controlling value of one rivet.

Problem 11-6-C.* In a joint similar to that shown in Fig. 11-4e, the two outer plates are $\frac{1}{4}$-in. thick and the enclosed plate is $\frac{5}{16}$ in. thick. If $\frac{3}{4}$-in. rivets are to be used, determine the controlling value of one rivet.

Problem 11-6-D.* Two outer $\frac{5}{16}$-in. plates enclose a $\frac{7}{16}$-in. plate in an arrangement similar to that shown in Fig. 11-4e. State whether the controlling value of one rivet is determined by shear or bearing if $\frac{7}{8}$-in. rivets are used.

11-7 Bolted Connections

Bearing-type connections may be made with bolts as well as rivets, but the current wide use of bolting in structural work has resulted from the development of the friction-type connection made possible by high-strength bolts.

Bolts are similar to rivets in that they consist of a head and a shank, but the bolt shank is threaded to receive a nut. As noted in Art. 3-6, the 1969 AISC Specification requires that material from which bolts are made conform to one of four ASTM specifications. Bolts are commonly designated by their ASTM specification number; the high-strength group consists of A325, A449, and A490 bolts; all others are denoted as A307 bolts. When speaking of bolts in general, the term "threaded fasteners" is often used in contradistinction to rivets. All bolts commonly used in steel building construction are classified as unfinished or high-strength.

11-8 Unfinished Bolts

Under certain conditions in bearing-type connections, unfinished bolts are used in place of rivets. These are made by automatic machines from rods of A307 steel as they come from the mill. The sizes of unfinished bolts are not exact, and the allowable working unit stresses are considerably lower that the allowable stresses for rivets. This difference is shown in Table 11-1, where the allowable unit shearing stress for A502, Grade 1 rivets is listed as 15 ksi and that for A307 bolts is listed as 10 ksi. The allowable unit bearing stress, however, is 48.6 ksi for both A307 bolts and rivets. Unfinished bolts are not permitted under some building codes in the construction of buildings exceeding certain height limits; nor should they be used for connections subjected to vibration, impact, or fatigue. Where these conditions exist, nuts may become loose, impairing the strength of the connection. The holes for unfinished bolts, as for rivets, should be $\frac{1}{16}$ in. larger than the nominal diameter of the bolt.

When using bolts in bearing-type connections, it is extremely important to see that the threaded portion of the shank does not bear on the metal being connected and that it does not occur at the shear planes. Such a condition will exist if the bolt is threaded for too great a distance from its nut end. If the bolt is too long, washers are used to give a full grip when the nut is turned tight.

Ribbed Bolts. Another type of bolt developed in recent years is the ribbed bolt. This bolt has on the shank longitudinal ribs or flutes which project from its core, giving an overall diameter somewhat larger than the diameter of the hole. These bolts on being driven into the holes are wedged tightly by the deformation of the ribs against the sides of the fitted pieces. They provide greater resistance to vibration than ordinary bolts. The threaded end of the bolt receives a locknut and washer; and when properly installed, the bolt is as strong as a rivet of the same diameter.

11-9 High-Strength Bolts

Figure 11-5*a* indicates two plates held together by a high-strength bolt with a nut and two washers. This is a friction-type connection. The nut is tightened to provide a tensile stress, T, in the bolt that approaches its yield stress, thus clamping the plates tightly together. The frictional resistance, S, prevents slip between the plates. When P, the joint load, exceeds the magnitude of S, slip occurs. If there is slip between the plates, the edges of the plates are brought in contact with the shank of the bolt, thereby producing bearing and shearing stresses in the bolt. If there is *no slip*, the load P is transmitted from one plate to the other by the *frictional* resistance that results from T, the high tension in the bolt. In this case, there are no shearing or bearing stresses on the bolt. Theoretically, the bolt's shank and the surfaces of the plates through which it passes are not in contact at all because the holes are punched slightly larger than the bolt. Even when a connection of this type is subjected to vibration, the high residual tensile stress prevents loosening.

Before tension is applied to high-strength bolts, the joint surfaces

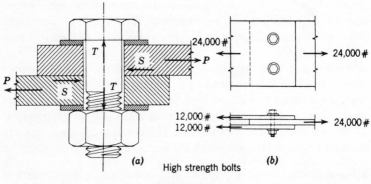

(a) High strength bolts   (b)

FIGURE 11-5

adjacent to bolt heads, nuts, or washers must be free of scale, burrs, dirt, and other foreign material. The tensile stresses set up in the bolts must be controlled; and when all fasteners in the joint are tight, the minimum bolt tension should approximate the values given in Table 11-3. Tightening of high-strength bolts in friction-type

TABLE 11-3. High-Strength Bolt
Tension

Nominal bolt diameter in inches	Minimum bolt tension (proof load) in kips
$\frac{5}{8}$	19.2
$\frac{3}{4}$	28.4
$\frac{7}{8}$	30.1
1	47.3

connections is accomplished by special torque or impact wrenches. Because the efficiency of this connection depends on the control used in achieving the proper tensile stresses in the bolts, special attention must be given to supervision of their installation and tightening.

High-strength bolts are usually made of A325 or A449 steel. The design procedure is predicated on taking the allowable working value of a bolt equal to the allowable shear value of a rivet of equivalent size, with no consideration of bearing stresses. In friction-type connections, this hypothetical shear value is based on an allowable shearing stress of 15 ksi whether or not threading is excluded from the shear plates (Table 11-1).

When high-strength bolts are used in connections where tightening of the nut is done by hand wrenches, resulting in only a snug fit, there is insufficient tension developed to prevent slip. The allowable working value of a bolt is then determined as for any bearing-type connection, the allowable shear stress for A325 bolts being limited to 15 ksi when threading is *not* excluded from the shear planes and 22 ksi when threading is excluded.

Example. Three plates are to be connected with high-strength, A325 bolts, as shown in Fig. 11-5. This is a friction-type connection.

The outer plates are each ¼ in. thick, and the enclosed plate is ½ in. thick. The load transferred from one side of the connection to the other is 24 kips. How many ¾-in. bolts are required?

Solution: (1) As already explained, high tensile stresses are set up in the bolts, and the clamping forces result in a frictional "shearing" resistance between the plates to be connected. The allowable load for one bolt is considered equal to the allowable shear value of a rivet of the same diameter; therefore, Table 11-2 may be used even though part of it relates to bearing-type connections.

(2) Because of the arrangement of the connection, we will use the double shear value given in Table 11-1 under the ¾-in. diameter heading. This is 13.25 kips.

(3) The number of bolts required = 24 ÷ 13.25 = 1.8 or 2 bolts. No consideration need be given to bearing strength in this friction-type connection.

11-10 Gage Lines

The lines parallel to the length of a member on which rivets or bolts are placed are called *gage lines*. The gage is the perpendicular distance between gage lines or between a gage line and the edge of a member. The usual gage dimensions for standard structural steel angles are given in Table 11-4. The AISC Manual gives, under dimensions for

TABLE 11-4. Usual Gage Dimensions for Angles*

Leg	8	7	6	5	4	3½	3	2½	2
g	4½	4	3½	3	2½	2	1¾	1⅜	1⅛
g_1	3	2½	2¼	2					
g_2	3	3	2½	1¾					

* Reproduced from data in the 7th Edition of the *Manual of Steel Construction*. Courtesy American Institute of Steel Construction.

detailing, the usual gage dimensions for the flanges of wide flange shapes, standard I-beams, and channels.

11-11 Pitch of Bolts and Rivets

The center-to-center distance between adjacent bolts or rivets, whether they are on the same or different gage lines, is called the *pitch*. The minimum distance between the centers of bolt and rivet holes is $2\frac{2}{3}$ times the nominal diameter of the fastener, but preferably the minimum pitch should not be less than 3 diameters. Table 11-5 gives the minimum pitch to maintain 3 diameters when

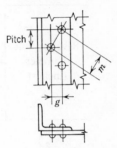

TABLE 11-5. Minimum Pitch to Maintain Three Diameters Center to Center of Holes*

Diameter of rivet	m	Distance, g, inches								
		1	$1\frac{1}{4}$	$1\frac{1}{2}$	$1\frac{3}{4}$	2	$2\frac{1}{4}$	$2\frac{1}{2}$	$2\frac{3}{4}$	3
$\frac{5}{8}$	$1\frac{7}{8}$	$1\frac{5}{8}$	$1\frac{3}{8}$	$1\frac{1}{8}$	$\frac{5}{8}$	0				
$\frac{3}{4}$	$2\frac{1}{4}$	2	$1\frac{7}{8}$	$1\frac{5}{8}$	$1\frac{3}{8}$	1	0			
$\frac{7}{8}$	$2\frac{5}{8}$	$2\frac{1}{2}$	$2\frac{3}{8}$	$2\frac{1}{8}$	2	$1\frac{3}{4}$	$1\frac{3}{8}$	$\frac{3}{4}$	0	
1	3	$2\frac{7}{8}$	$2\frac{3}{4}$	$2\frac{5}{8}$	$2\frac{1}{2}$	$2\frac{1}{4}$	2	$1\frac{1}{8}$	$1\frac{1}{8}$	0

* Reproduced from data in the 7th Edition of the *Manual of Steel Construction*. Courtesy American Institute of Steel Construction.

the holes in adjacent rows are staggered. Sometimes the physical arrangement of a connection requires a plate that will accommodate more bolts or rivets that the minimum necessary; in such cases, when the stresses on the fasteners are relatively small, 6 in. is considered to be the maximum pitch.

11-12 Edge Distance

If a bolt or rivet is placed too close to the edge of a member, there is a tendency to tear or deform the adjacent metal. To overcome this type of failure, certain prescribed distances are maintained between the center of the hole and the edge of the member. This is called the *edge distance*. Greater edge distances are required for sheared edges than for the edges of rolled sections and plates. The AISC Specification requires that the distance from the center of a rivet or bolt hole to the edge of any member shall not be less than that given in Table 11-6.

TABLE 11-6. Minimum Edge Distances for Holes*

Rivet or bolt diameter (inches)	Minimum edge distance for punched, reamed, or drilled holes (inches)	
	At sheared edges	At rolled edges of plates, shapes, or bars or gas cut edges†
⅝	1⅛	⅞
¾	1¼	1
⅞	1½**	1⅛
1	1¾**	1¼

* Reproduced from data in the 7th Edition of the *Manual of Steel Construction*. Courtesy American Institute of Steel Construction.
** These may be 1¼ in. at the ends of beam connection angles.
† All edge distances in this column may be reduced ⅛ in. when the hole is at a point where stress does not exceed 25% of the maximum allowed stress in the element.

The maximum distance from the center of any rivet or bolt to the nearest edge of a plate shall be 12 times the thickness of the plate but not over 6 in.

11-13 Net Sections

As noted in Art. 11-3, a possible type of failure of a riveted joint is by tension in the net section in one of the connected members. Net section is defined as the gross cross-sectional area of a member minus holes or openings in the plane of the normal cross section. As stated earlier, rivet and bolt holes are punched $\frac{1}{16}$ in. larger in diameter than the nominal diameter of the fastener. The punching damages a small amount of the steel around the perimeter of the

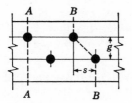

FIGURE 11-6

hole, and consequently, the diameter of the hole to be deducted in determining the net section is ⅛ in. greater than the nominal diameter of the rivet.

Where only one hole is involved, as in Fig. 11-3c, or in a similar connection with a single row of fasteners along the line of stress, the net area of the cross section of one of the plates is found by multiplying the plate thickness by its net width (width of member minus diameter of hole).

Where holes are staggered in two rows along the line of stress (Fig. 11-6), the net section is determined somewhat differently. The AISC Specification reads:

In the case of a chain of holes extending across a part in any diagonal or zigzag line, the net width of the part shall be obtained by deducting from the gross width the sum of the diameters of all the holes in the chain and adding, for each gage space in the chain, the quantity $s^2/4g$, where

s = longitudinal spacing (pitch) in inches of any two successive holes

and

g = transverse spacing (gage) in inches for the same two holes

The critical net section of the part is obtained from that chain which gives the least net width.

The AISC Specification also provides that in no case shall the net section through a hole be considered as more than 85% of the corresponding gross section.

Example. The plate shown in Fig. 11-6 is 5 in. wide and ½ in. thick. The gage is 1¾ in., the pitch is 2 in., and the rivets to be used are ⅞ in. Determine the net section.

Solution: (1) The diameter of one hole is $\frac{7}{8} + \frac{1}{8} = 1$ in. Therefore, the net width along the line A-A is $5 - 1 = 4$ in. As the thickness of the plate is $\frac{1}{2}$ in., the net area at section A-A is $4 \times \frac{1}{2} = 2$ sq in.

(2) In accordance with the foregoing specification, the net width on the broken line B-B is

$$5 - 2 + \frac{s^2}{4g} = 5 - 2 + \frac{2^2}{4 \times 1.75} = 3.57 \text{ in.}$$

and the net area on line B-B is $3.57 \times 0.5 = 1.78$ sq in.

(3) The critical net section of the plate is the lesser of the two values found above, or 1.78 sq in.

Tension Members.   All members in direct tension are designed on the basis of the net section. Single and double angles are often used as hangers, and double angles are widely used as tension members of roof trusses. The design of tension members for roof trusses is discussed in Art. 13-17.

11-14 Framing Connection Details

When one beam is supported by another by being placed above it, as indicated in Fig. 11-7a, bolts or rivets are used only to hold the beams in position. A framed connection using connecting angles is shown in Fig. 11-7b. This type of connection is the most common. It is used for framing beams to girders but less frequently for framing beams to columns. A seated connection formed with an unstiffened shelf angle is shown in Fig. 11-7c and d. For this connection, a side angle is required to hold the beam in position; it is assumed that the entire load from the beam is transferred to the girder web by the rivets in the shelf angle. Although rivets are indicated in the figures,

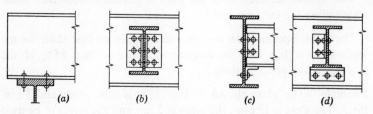

FIGURE 11-7

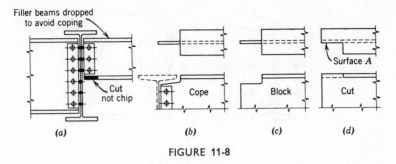

FIGURE 11-8

these connections may be made with bolts also. The use of welds in framing connections is discussed in Chapter 12.

When two beams frame together so that the upper surfaces of the top flanges are at the same level, the term *flush top* is used. To accomplish this, it is necessary to cut away a portion of the upper flange (Fig. 11-8b and c). This is known as *coping* or *blocking* and for economy should be avoided if possible. When the type of floor construction supported by the floor beams will permit, the elevations of the tops of the beams should be established a sufficient distance below the tops of the girders to clear the girder fillet. This arrangement is shown in Fig. 11-8a.

When the web of a spandrel beam frames against the face of a column, it becomes necessary to cut away a portion of one side of the flange, as indicated in Fig. 11-8d. The flange surface marked A is not flush with the web unless it is chipped. When making detail drawings, it is necessary to specify by note "cut and chip" or "cut not chip," whichever is desired.

11-15 Beam Connections to Girders

Figure 11-9 shows in more detail the type of connection illustrated in Fig. 11-7b and discussed in the preceding article. The usual method of attaching the angles to the beam web when rivets are used is indicated in Fig. 11-9b and c. The girder web, against which the outstanding legs of the angles frame, does not appear in the illustration. The method of determining the strength of such a connection is given in the following example.

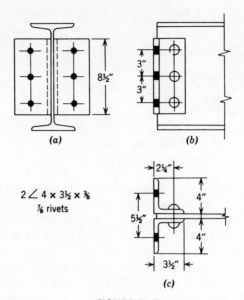

(a)   (b)

$2 \angle 4 \times 3\frac{1}{2} \times \frac{3}{8}$
$\frac{7}{8}$ rivets

(c)

FIGURE 11-9

Example 1.   The rivets are $\frac{7}{8}$ in. in diameter, and the beam to which the angles are attached is an S 12 × 31.8. Table 1-2 lists the web thickness of the beam as 0.350 in. Determine the magnitude of the greatest beam reaction this connection will support if the angles are $\frac{3}{8}$ in. thick.

Solution: (1) First consider the three $\frac{7}{8}$-in. rivets which connect the two angles to the web of the standard I-beam. These rivets are in double shear. Referring to Table 11-2, we find that the double shear rivet value is 18.04 kips. Since there are three rivets, their total capacity is 3 × 18.04 = 54.1 kips.

(2) The bearing capacity will be determined by the bearing area of the web of the S 12 × 31.8 or the combined thickness of the two angles ($\frac{3}{8} + \frac{3}{8} = \frac{3}{4}$ or 0.75 in.), whichever is thinner. Since the web thickness is 0.350 in., it will control and the bearing value of one rivet becomes

$$\text{Rivet value (0.35-in. bearing)} = F_p \times A$$
$$= 48.6 \times 0.875 \times 0.350$$
$$= 14.9 \text{ kips}$$

There are three rivets, so the capacity of the connection as governed by bearing is $3 \times 14.9 = 44.7$ kips.

(3) Of the two values computed above, 44.7 kips is the smaller. Therefore, the greatest permissible beam reaction for the connection is 44.7 kips, provided the outstanding legs of the angles can develop this value in the girder web.

(4) Assume that the supporting girder is a W 18 × 50 with a web thickness of 0.358 in. (Table 1-1). If there is *not* a beam framing into the other side of the girder opposite this one, the rivets in the outstanding legs of the connection angles are in single shear, and the capacity of one rivet is 9.02 kips (Table 11-2).

The bearing capacity of one rivet will be controlled by the thickness of the girder web (0.358 in.) or the thickness of one angle ($\frac{3}{8} = 0.375$ in.), whichever is the smaller. Then

$$\text{Rivet value (0.358-in. bearing)} = F_p \times A$$
$$= 48.6 \times 0.875 \times 0.358$$
$$= 15.2 \text{ kips}$$

This is larger than the single shear capacity of one rivet, so the controlling value is 9.02 kips, making the capacity of the group $6 \times 9.02 = 54.1$ kips. Therefore, the maximum reaction the connection will support is the 44.7 kips determined in step 3, above.

Example 2. In the preceding example, it was found that the greatest permissible beam reaction for the connection shown in Fig. 11-9 was 44.7 kips, controlled by bearing in the web of the S 12 × 31.8. If another S 12 × 31.8 frames into the other side of the W 18 × 50 girder directly opposite, determine whether the maximum reaction *each* beam can transmit to the girder is still 44.7 kips.

Solution: (1) The rivets that pierce the girder web are now in double shear, and the capacity of one rivet is 18.04 kips (Table 11-2).

(2) Since the combined thickness of the two angles is 0.75 in. and the web thickness of the girder is 0.358 in., the working value for one rivet is again controlled by bearing in the girder web. This was found in step 4 of Example 1 to be 15.2 kips.

(3) The total load that can be transferred to the girder through this connection is $6 \times 15.2 = 91.2$ kips. Therefore, 44.7 kips remains the maximum reaction each S 12 × 31.8 can transmit to the girder ($44.7 + 44.7 = 89.4 < 91.2$).

11-16 Framed Beam Connections

It is not necessary for the designer to compute the proper size of connecting angles and the number of bolts or rivets for the usual framed connection. Standard beam connections have been developed and tables prepared to facilitate their use. Table 11-7, abstracted from the 7th Edition of the AISC Manual, illustrates the makeup of such tables. It lists shear capacities, bearing capacities, and angle sizes for framed beam connections composed of three and four rows of fasteners, either bolts or rivets. This table can be used for shear capacities of $\frac{3}{4}$-in., $\frac{7}{8}$-in., and 1-in. A307 bolts, A502-Grade 1 and 2

TABLE 11-7. Framed Beam Connections Bolted or Riveted, Allowable Loads in kips**

4 Rows|

W 24, 21, 18, 16
S 24, 20, 18, 15
C 18, 15

Varies

$\frac{3}{4}''\phi, \frac{7}{8}''\phi$ — 2¼
$1''\phi$ — 2½

TABLE I-A4 Total Shear, kips

^{a}Fastener designation	F_v ksi	Fastener diameter					
		$\frac{3}{4}$		$\frac{7}{8}$		1	
		Load	t^b	Load	t^b	Load	t^b
A307	10.0	35.4	¼	48.1	¼	62.8	¼
A325-F A325-N A502-1	15.0	53.0	¼	72.2	¼	94.2	$\frac{5}{16}$
A490-F A502-2	20.0	70.7	¼	96.2	$\frac{5}{16}$	126	$\frac{7}{16}$
A325-X	22.0	77.8	$\frac{5}{16}$	106	$\frac{3}{8}$	138	$\frac{7}{16}$
A490-N	22.5	79.5	$\frac{5}{16}$	108	$\frac{3}{8}$	141	$\frac{7}{16}$
A490-X	32.0	113	$\frac{7}{16}$	154	¼	201	$\frac{5}{8}$

TABLE 1-B4. Total Bearing,^{c} in ksi, 4 Fasteners on 1″ Thick Material

	F_y	36	42	45	50	55	60	65	100
Fastener diameter	$\frac{3}{4}$	146	170	182	203	223	243	263	405
	$\frac{7}{8}$	170	198	213	236	260	284	307	473
	1	194	227	243	270	297	324	351	540

** Taken from a more complete set of tables in the 7th Edition of the *Manual of Steel Construction*. Courtesy American Institute of Steel Construction.
$_a$ For description of fastener designation see page 4-12.

rivets, and A325 and A490 high-strength steel bolts. Bearing capacities for the three fastener diameters are given for a *1-in. thickness of material* for computation of connection values in steel of the yield points listed. The description of fastener designation cited in footnote *a* to the table is given below.

A325-F and A490-F: Friction-type connection
A325-N and A490-N: Bearing-type connections with threads included in shear plane
A325-X and A490-X: Bearing-type connections with threads excluded from shear plane

3 Rows		

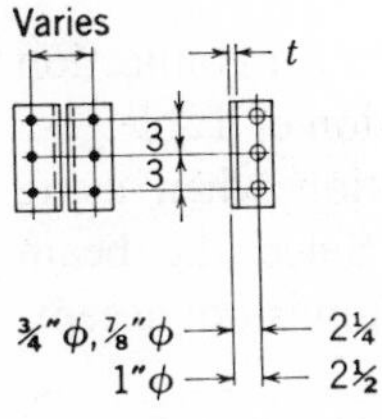

TABLE 1-A3. Total Shear, kips

W 18, 16, 14, 12, 10*
S 18, 15, 20, 10*
C 18, 15, 13, 12, 10*

Varies

¾″φ, ⅞″φ → |← 2¼
1″φ → |← 2½

ᵃFastener designation	F_v ksi	Fastener diameter					
		¾		⅞		1	
		Load	t^b	Load	t^b	Load	t^b
A307	10.0	26.5	¼	36.1	¼	47.1	¼
A325-F A325-N A502-1	15.0	39.8	¼	54.1	¼	70.7	⁵⁄₁₆
A490-F A502-2	20.0	53.0	¼	72.2	⁵⁄₁₆	94.3	⁷⁄₁₆
A325-X	22.0	58.3	⁵⁄₁₆	79.4	⅜	104	⁷⁄₁₆
A490-N	22.5	59.6	⁵⁄₁₆	81.2	⅜	106	⁷⁄₁₆
A490-X	32.0	84.8	⁷⁄₁₆	115	½	151	⅝

TABLE 1-B3. Total Bearing,ᶜ in kips, 3 Fasteners on 1″ Thick Material

	F_y	36	42	45	50	55	60	65	100
Fastener diameter	¾	109	128	137	152	167	182	197	304
	⅞	128	149	159	177	195	213	230	354
	1	146	170	182	203	223	243	263	405

ᵇ Thickness *t* based on connection angles of $F_y = 36$ ksi material.
ᶜ Use decimal thickness of enclosed web material as multiplying factor for these values.
* Limited to W 10 × 11.5, 15, 17, 19, 21, 25, 29; S 10 × 9; C 10 × 6.5, 8.4.

These are for high-strength bolts; the designations for A307 bolts and A502 rivets are clear from the table.

It will be noted that Table 11-7 is divided into two parts, for connections with three and four rows of fasteners. In each part, there is a portion marked "Table A" in which the shear values for fasteners of the three dimensions and different grades of steel are listed. In this part of the table, t, the thickness of the connection angles, is given. Use of the bearing values given in the portion marked "Table B" is explained in the examples given below.

In order that the most economical connections can be provided by the steel fabricator, the magnitude of beam reactions should be shown on the structural working drawings. When beam reactions are known, the series of tables from which Table 11-7 is taken are highly effective in determining the proper connections.

Example 1. A W 18 × 50 of A36 steel has a web thickness of 0.358 in. The rivets in the connection are ⅞ in. in diameter and are made of A502-1 steel. If the beam reaction to be transmitted is 55 kips, determine by the use of Table 11-7 the makeup of an appropriate connection.

Solution: (1) Since this is an 18-in. beam, let us try a connection using three rows of rivets. Referring to the "A" portion of Table 11-7 for three rows, the load capacity of the connection when using ⅞-in. A502-1 rivets is found to be 54.1 kips. Since the beam reaction is 55 kips, the three-rows-of-rivets connection is not acceptable.

(2) Moving to the four-row portion of the table, we find that the load capacity of the connection using ⅞-in. rivets of A502-1 steel is 72.2 kips. This exceeds the magnitude of the beam reaction and consequently is acceptable as the shear value of the connection.

(3) To investigate the bearing capacity of the 4 rivets as controlled by the enclosed web of the beam, we use the "B" portion of the table. The value of 170 kips shown for ⅞-in. fastener diameter and bearing in a steel with a yield point of 36 ksi is for a plate *1 in. thick.* To find the capacity for the 0.358-in. thickness of the web of the W 18 × 50, we multiply the tabular value of 170 kips by this decimal thickness. Then 170 × 0.358 = 60.8 kips.

(4) Since both 72.2 kips and 60.8 kips exceed the beam reaction of 55 kips, the connection composed of two angles ¼ in. thick with 4 rows of rivets is acceptable.

Example 2. By the use of Table 11-7, determine the makeup of a connection for a W 12 × 31 of A36 steel. The beam reaction is 20 kips, and ¾-in., A325 high-strength bolts with threads included in the shear plane are to be used in this bearing-type connection.

Solution: (1) In view of the 12-in. depth of the beam, we will try a connection with three rows of rivets. The "A" portion of Table 11-7 shows the shear capacity of 39.8 kips for a 3-row connection using ¾-in., A325 bolts with threads included in the shear plane (A325-N).

(2) The web thickness of the W 12 × 31 is given in Table 1-1 as 0.265 in. From the "B" portion of Table 11-7, the bearing capacity of the connection in 1-in. thick metal is found to be 109 kips for ¾-in. diameter fasteners on A36 steel. Therefore, the bearing capacity of the connection as controlled by the thickness of the beam web is 109 × 0.265 = 28.9 kips.

(3) Since both the shear and bearing capacities (39.8 kips and 28.9 kips) exceed the beam reaction of 20 kips, the connection composed of two angles ¼-in. thick with three rows of bolts is acceptable.

Problem 11-16-A. With the aid of Table 11-7, verify that the beam connection shown in Fig. 11-9 with ⅞-in. rivets and an S 12 × 31.8 will support a reaction of 44.7 kips.

11-17 Beam Connections to Columns

Because the connection of beams to columns presents a great variety of conditions, it is not practicable to use uniform connections to the extent possible with beam-to-girder connections. Some typical beam-to-column connections are shown in Fig. 11-10. Although rivets are indicated on the sketches, bolts are extensively employed in a similar manner. Welded beam-to-column connections are discussed in Chapter 12.

Seated connections with stiffeners are commonly used for the larger beams. The stiffened seat connections illustrated in Fig. 11-10a, b, and c consist of a shelf angle, filler, and single or double

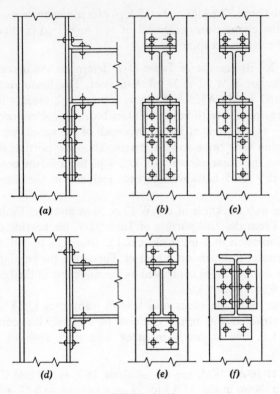

FIGURE 11-10

stiffener angles. The stiffener angles must fit tightly against the outstanding leg of the shelf angle. The bolts or rivets penetrating the stiffener leg and column flange are in single shear and are computed to transmit the entire beam reaction to the column. The filler has, of course, the same thickness as the shelf angle. The *clip angle* at the top is used merely to hold the beam in position and does not help transfer the beam load to the column. If preferred, the clip angle shown at the top in Fig. 11-10*a*, *b*, and *c* may be placed against the web of the beam, similar to the position indicated in Fig. 11-7*d*.

Seat connections without stiffeners (Fig. 11-10*d* and *e*) may be used for beams with smaller reactions. The strength of this type of connection depends on the size and thickness of the shelf angle and the number of bolts or rivets connecting it to the column. Unlike the stiffened seat connections, the shelf angle in Fig. 11-10*d* and *e* must resist bending stresses created by the tendency of the beam to deflect the outstanding leg of the seat. For this reason, stiffened beam seats are required for the larger beams when the reactions are greater. Framed connections (Fig. 11-10*f*) consisting of two angles riveted to the web of the beam are similar to connections for framing beams to girders. The entire beam reaction is transferred to the column by the fasteners in these connection angles. The shelf angle below the beam flange is used merely as an aid in erection.

In addition to the series of tables similar to Table 11-1, the AISC Manual contains many tables to assist in the design of both stiffened and unstiffened seat connections, together with examples illustrating their use. The reader interested in pursuing the subject of bolted and riveted connection design further should consult Part 4 of the Manual.

11-18 Conventional and Moment Connections

All of the beam-to-girder and beam-to-column connections discussed in this chapter come under the category of "simple" or free-end connections. That is, insofar as gravity loading is concerned, the ends of the beams and girders are connected for shear only and are free to rotate under gravity load. We shall call connections of this nature *conventional connections;* they are used in Type 2 of the three types of steel construction recognized in the AISC Specification.

Type 1 construction, commonly designated as continuous or rigid frame, assumes that beam-to-column connections possess sufficient rigidity to prevent rotation of the beam ends as the member deflects under its load. This means that the connection must transmit some bending moment between beam and column, and consequently it is called a moment-resisting or simply a *moment connection.*[1] Although Type 1 continuous framing can be achieved by proper

---

[1] Type 3 construction, called partially restrained or semi-rigid framing, assumes that the connections possess a dependable and known moment-resisting capacity of a degree between the rigidity of Type 1 and the flexibility of Type 2.

design of bolted or riveted connections, it is accomplished much more effectively in welded construction. This aspect of welded connections is considered briefly in Chapter 12. A fully continuous frame of Type 1 construction is statically indeterminate, and its analysis and design are beyond the scope of this book. Moment-resisting connections are used in multistory steel frame buildings to provide lateral stability against the effects of wind and earthquake forces.

In general, the design methods and procedures treated in this volume are applicable to Type 2 construction and follow the provisions of Section 1.12.1 of the AISC Specification, which states that "Beams, girders, and trusses shall ordinarily be designed on the basis of simple spans whose effective length is equal to the distance between centers of gravity of the members to which they deliver their end reactions."

12

Welded
Connections

IIIIIIIIIIIIIIIIIIIIIIIIIIIIIIIIIIIIIIIIIIIIIIIII

12-1 General

One of the distinguishing characteristics of welded construction is
the facility with which one member may be attached directly to
another without the use of additional plates or angles, which are
necessary in bolted and riveted connections. A welded connection
requires no holes for fasteners, and therefore the gross rather than
the net section may be considered when determining the effective
cross-sectional area of members in tension.

As noted in Art. 11-18, moment-resisting connections are readily
achieved by welding; consequently, welded connections are customary in Type 1 construction in order to develop continuity in the
framing. Welding may also be used in Type 2 construction, but care
must be exercised in design to assure that a rigid connection is not
provided where free-end conditions have been assumed in design of
the framing.

Welding is often used in combination with bolting in "shop-
welded and field-bolted construction." Here connection angles with
holes in the outstanding legs may be welded to a beam in the
fabricating shop and then bolted to a girder or column in the field.
Figure 12-10a shows such an arrangement.

12-2 Electric Arc Welding

Although there are many welding processes, electric arc welding is the one generally used in steel building construction. In this type of welding, an electric arc is formed between an electrode and the two pieces of metal that are to be joined. The intense heat melts a small portion of the members to be joined as well as the end of the electrode or metallic wire. The term *penetration* is used to indicate the depth from the original surface of the base metal to the point at which fusion ceases. The globules of melted metal from the electrode flow into the molten seat and, when cool, are united with the members that are to be welded together. *Partial penetration* is the failure of the weld metal and base metal to fuse at the root of a weld. It may result from a number of items, and such incomplete fusion produces welds that are inferior to those of full penetration.

12-3 Welded Joints

When two members are to be joined, the ends may or may not be grooved in preparation for welding. In general, there are three classification of joints: *butt joints*, *tee joints*, and *lap joints*. The selection of the type of weld to use depends on the magnitude of the load requirement, the manner in which it is applied, and the cost of

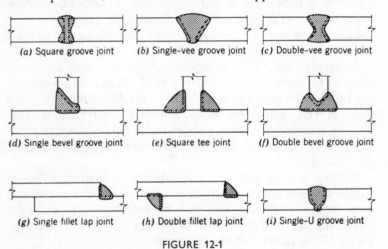

(a) Square groove joint   (b) Single-vee groove joint   (c) Double-vee groove joint

(d) Single bevel groove joint   (e) Square tee joint   (f) Double bevel groove joint

(g) Single fillet lap joint   (h) Double fillet lap joint   (i) Single-U groove joint

FIGURE 12-1

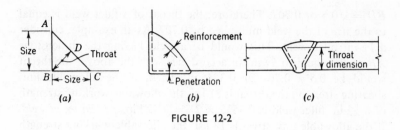

FIGURE 12-2

preparation and welding. Several joints are shown in Fig. 12-1. The type of joint and preparation permit a number of variations. In addition, welding may be done from one or both sides. The scope of this book prevents a detailed discussion of all the many joints and their uses and limitations.

The weld most commonly used for structural steel in building construction is the *fillet weld*. It is approximately triangular in cross section and is formed between the two intersecting surfaces of the joined members. See Fig. 12-2a and b. The *size* of a fillet weld is the leg length of the largest inscribed isosceles right triangle, *AB* or *BC*. See Fig. 12-2a. The *root* of the weld is the point at the bottom of the weld, point *B* in Fig. 12-2a. The *throat* of a fillet weld is the distance from the root to the hypotenuse of the largest isosceles right triangle that can be inscribed within the weld cross section, distance *BD* in Fig. 12-2a. The exposed surface of a weld is not the plane surface indicated in Fig. 12-2a but is usually somewhat convex, as shown in Fig. 12-2b. Therefore, the actual throat may be greater than that shown in Fig. 12-2a. This additional material is called *reinforcement*. It is not included in determining the strength of a weld.

A single-vee groove weld between two members of unequal thickness is shown in Fig. 12-2c. The *size* of a butt weld is the thickness of the thinner part joined, with no allowance made for the weld reinforcement.

12-4   Stresses in Welds

If the dimension (size) of *AB* in Fig. 12-2a is 1 unit in length, $(AD)^2 + (BD)^2 = 1^2$. Since AD and BD are equal, $2(BD)^2 = 1^2$ and

$BD = \sqrt{0.5}$ or 0.707. Therefore, the throat of a fillet weld is equal to the *size* of the weld multiplied by 0.707. As an example, consider a $\frac{1}{2}$-in. fillet weld. This would be a weld having the dimensions *AB* or *BC* equal to $\frac{1}{2}$ in. In accordance with the above, the throat would be 0.5×0.707 or 0.3535 in. Then, if the allowable unit shearing stress on the throat is 21 ksi, the allowable working strength of a $\frac{1}{2}$-in. fillet weld is $0.3535 \times 21 = 7.42$ kips *per lin in. of weld.* If the allowable unit stress is 18 ksi, the allowable working strength is $0.3535 \times 18 = 6.36$ kips *per lin in. of weld.*

The permissible unit stresses used in the preceding paragraph are for welds made with E 70 XX and E 60 XX type electrodes on A36 Steel. Particular attention is called to the fact that *the stress in a fillet weld is considered as shear on the throat, regardless of the direction of the applied load.* Neither plug nor slot welds shall be assigned any values in resistances other than shear. The allowable working strengths of fillet welds of various sizes are given in Table 12-1, with values rounded to $\frac{1}{10}$ of a kip.

TABLE 12-1. Allowable Working Strength of Fillet Welds

Size of fillet weld (inches)	Allowable loads in kips per lin in.	
	E 60 XX electrodes $F_{vw} = 18$ ksi	E 70 XX electrodes $F_{vw} = 21$ ksi
$\frac{3}{16}$	2.4	2.8
$\frac{1}{4}$	3.2	3.7
$\frac{5}{16}$	4.0	4.6
$\frac{3}{8}$	4.8	5.6
$\frac{1}{2}$	6.4	7.4
$\frac{5}{8}$	8.0	9.3
$\frac{3}{4}$	9.5	11.1
1	12.7	14.8

The stresses allowed for the metal of the connected parts (known as the *base metal*) apply to complete penetration groove welds stressed in tension and compression parallel to the axis of the weld and in tension normal to the effective throat. They apply also to complete or partial penetration groove welds stressed in compression normal to the effective throat and in shear on the effective throat. Consequently, allowable stresses for butt welds are the same as for the base metal.

The relation between the weld size and the maximum thickness of material in joints connected only by fillet welds is shown in Table 12-2. The maximum size of a fillet weld applied to a square edge of a

TABLE 12-2. Relation between Material Thickness and Minimum Size of Fillet Welds*

Material thickness of thicker part joined (inches)	Minimum size of fillet weld (inches)	Material thickness of thicker part joined (inches)	Minimum size of fillet weld (inches)
To $\frac{1}{4}$ inclusive	$\frac{1}{8}$	Over $1\frac{1}{2}$ to $2\frac{1}{4}$	$\frac{3}{8}$
Over $\frac{1}{4}$ to $\frac{1}{2}$	$\frac{3}{16}$	Over $2\frac{1}{4}$ to 6	$\frac{1}{2}$
Over $\frac{1}{2}$ to $\frac{3}{4}$	$\frac{1}{4}$	Over 6	$\frac{5}{8}$
Over $\frac{3}{4}$ to $1\frac{1}{2}$	$\frac{5}{16}$		

* Taken from the 7th Edition of the *Manual of Steel Construction.* Courtesy American Institute of Steel Construction.

plate or section $\frac{1}{4}$ in. or more in thickness should be $\frac{1}{16}$ in. less than the nominal thickness of the edge. Along edges of material less than $\frac{1}{4}$ in. thick, the maximum size may be equal to the thickness of the material.

The effective area of butt and fillet welds is considered to be the effective length of the weld multiplied by the effective throat thickness. The minimum effective length of a fillet weld should not be less than four times the weld size. For starting and stopping the arc, approximately $\frac{1}{4}$ in. should be added to the design length of fillet welds.

Figure 12-3a represents two plates connected by fillet welds. The welds marked *A* are longitudinal; *B* indicates a transverse weld. If a load is applied in the direction shown by the arrow, the stress distribution in the longitudinal weld is not uniform, and the stress in the transverse weld is approximately 30% higher per unit of length.

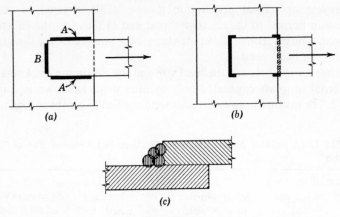

FIGURE 12-3

Added strength is given to a transverse fillet weld that terminates at the end of a member, as shown in Fig. 12-3b, if the weld is returned around the corner for a distance not less than twice the weld size. These end returns, sometimes called *boxing*, afford considerable resistance to the tendency of tearing action on the weld.

The ¼-in. fillet weld is considered to be the minimum practical size, and a 5⁄16-in. weld is probably the most economical size that can be obtained by one pass of the electrode. A small-size continuous weld is generally more economical than a larger discontinuous weld if both are made in one pass. Some specifications limit the single-pass fillet weld to 5⁄16 in. Large-size fillet welds require two or more passes (multipass welds) of the electrode, as shown in Fig. 12-3c.

Example.   A W 12 × 27 of A36 steel is to be welded to the face of a steel column with E 70 XX electrodes. See Fig. 12-4a. With respect

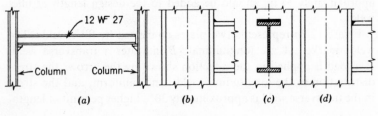

FIGURE 12-4

to the upper flange only, compute the strength of fillet and butt welds. Assume the beam is to be welded to produce continuous action as in Type 1 construction (Art. 11-18) so the upper flange will be in tension at the column. See Fig. 12-7a.

Solution: (1) First assume that the left end of the beam is in contact with the column and that a ⅜-in. fillet weld is run across the upper face of the beam flange as shown in Fig. 12-4b and c. Table 1-1 shows that the flange width of the W 12 × 27 is 6.5 in. and the flange thickness is 0.40 in. The length of the fillet weld, therefore, is 6.5 in.

(2) Referring to Table 12-1 under E 70 XX electrodes, the strength of a ⅜-in. fillet weld is given as 5.6 kips per linear inch, making the allowable strength of the weld 5.6 × 6.5 = 36.4 kips. Note that this weld resists tensile stresses, but as stated earlier, the strength of a fillet weld is determined by shear at the throat of the weld.

(3) Next, instead of the fillet weld, suppose that the upper flange is beveled and that a butt weld is used, as shown in Fig. 12-4d. The area of the weld resisting tension is the flange width multiplied by the flange thickness, or 6.5 × 0.40 = 2.6 sq in. Since the allowable tensile stress for the A36 steel from which the beam is made is 22 ksi (Table 3-2), the allowable strength of the butt weld in tension is 22 × 2.6 = 57.2 kips.

Problem 12-4-A A W 14 × 43 of A36 steel is to be welded with E 70XX electrodes to the fact of a column to produce continuous action. With respect to the upper flange only, determine the working strength of (a) a ½-in. fillet weld and (b) a butt weld as given in the foregoing example.

12-5 Design of Welded Joints

The most economical choice of weld to use for a given condition depends on several factors. It should be borne in mind that members to be connected by welding must be firmly clamped or held rigidly in position during the welding process. When riveting a beam to a column, it is necessary to provide a seat angle as a support to keep the beam in position for riveting the connecting angles. The seat angle is not considered as adding strength to the connection. Similarly, with welded connections, seat angles are commonly used. The designer must have in mind the actual conditions during erection and must provide for economy and ease in working the welds. Seat angles or similar members used to facilitate erection are *shop-welded* before the material is sent to the site. The welding done during erection is called *field welding*. The designer in preparing welding details indicates shop or field welds on the drawings. Conventional welding symbols are used to indicate the type, size, and position of the various welds. Only engineers or architects experienced in the design of welded connections should design or supervise welded construction. It is apparent that a wide variety of connections is possible; experience is the best aid in determining the most economical and practical connection.

The following examples illustrate the basic principles on which welded connections are designed.

Example 1. A bar of A36 steel, 3 × ⁷⁄₁₆ in. in cross section, is to be welded with E 70 XX electrodes to the back of a channel so that the full tensile strength of the bar may be developed. What should be the size of the weld?

Solution: (1) The area of the given bar is found to be 1.313 sq in. Since the allowable unit tensile stress of the steel is 22 ksi (Table 3-2), the tensile strength of the bar is $F_t \times A = 22 \times 1.313 = 28.9$ kips. The weld, therefore, must be of ample dimensions to resist a force of this magnitude.

(2) The bar being ⁷⁄₁₆ in. thick, a ³⁄₈-in. fillet weld will be used. Table 12-1 gives the allowable working strength as 5.6 kips per lin in. Hence the required length of ³⁄₈-in. fillet weld to develop the strength of the bar is $28.9 \div 5.6 = 5.16$ in. The position of the weld with respect to the bar is optional, and various arrangements are shown in Fig. 12-5a, c, and d.

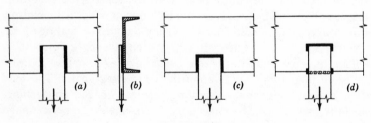

FIGURE 12-5

Example 2.   A $3\frac{1}{2} \times 3\frac{1}{2} \times \frac{5}{16}$-in. angle of A36 steel subjected to a tensile load is to be connected to a plate by means of fillet welds, using E 70 XX electrodes. What should the dimensions of the welds be to develop the full tensile strength of the angle?

Solution: (1) We shall use a $\frac{1}{4}$-in. fillet weld which has an allowable working strength of 3.7 kips per lin in. (Table 12-1). The cross-sectional area of the angle is 2.09 sq in. (Table 1-4). Since A36 structural steel has an allowable unit tensile stress of 22 ksi (Table 3-2), the tensile strength of the angle is $22 \times 2.09 = 46$ kips. Therefore, the required length of $\frac{1}{4}$-in. fillet welds to develop the strength of the angle is $46 \div 3.7 = 12.4$ in.

(2) An angle is an unsymmetrical cross section and the welds marked L_1 and L_2 in Fig. 12-6 will be made unequal in length so that their stresses will be proportioned in accordance with the distributed area of the angle. Referring to Table 1-4, we find that the centroid of the angle section is 0.99 in. from the back of the angle; hence, the two welds are 0.99 and 2.51 in. from the centroidal axis, as shown in Fig. 12-6. The lengths of welds L_1 and L_2 are made inversely proportional to their distances from the axis, but the sum of their length is

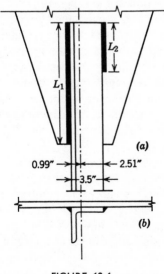

FIGURE 12-6

12.4 in. Therefore,

$$L_1 = \frac{2.51}{3.5} \times 12.4 = 8.9 \text{ in.}$$

and

$$L_2 = \frac{0.99}{3.5} \times 12.4 = 3.5 \text{ in.}$$

These are the design lengths required; and as noted earlier, each weld would actually be made $\frac{1}{4}$ in. longer than its required design length.

Problem 12-5-A.* A $4 \times 4 \times \frac{1}{2}$-in. angle of A36 steel is to be welded to a plate with E 70 XX electrodes to resist the full tensile strength of the angle. Using $\frac{3}{8}$-in. fillet welds, compute the design lengths L_1 and L_2, as shown in Fig. 12-6.

12-6 Beams with Continuous Action

As noted earlier, one of the advantages of welding is that beams having continuous action at the supports (Type 1 construction) are readily provided for. The usual bolted or riveted connections of Type 2 construction are assumed to offer no rigidity at the supports, and the bending moment throughout the length of the beam is positive. By the use of welding, however, a beam may be connected at its supports in such a manner that the beam is *fixed* or *restrained*, and a negative bending moment results (Art. 4-12). For the same span and loading, the maximum bending moment for a continuous beam is smaller than for a simple beam, and a lighter beam section is required.

When beams are rigidly connected by means of moment-resisting connections, the fibers in the upper flange *at the supports* are in tension, and the lower flange is in compression. This is shown diagrammatically in Fig. 12-7a. Therefore, in designing the welds for beams which have continuous action, we must provide for tension and compression in the upper and lower flanges, respectively, at the supports. A wide variety of welds is possible, and the following example illustrates the principles by which they are designed.

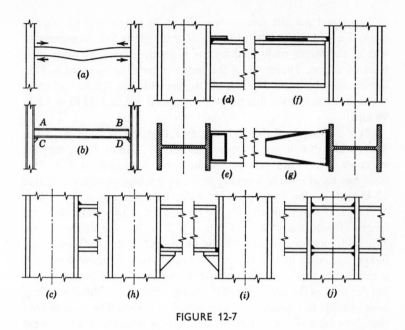

FIGURE 12-7

Example. A W 12 × 40 framing into column flanges at its ends is to be connected by welding so that a continuous action results. For erection of the beam, it is necessary that its length be slightly shorter than the distance between the flanges of the columns to which it will be welded. Seat angles are shop-welded to the columns, and the beam is supported on these during field welding of the connection. For this example, consider the left end of the beam to be tightly held against the column flange; this leaves a short space between the right end of the beam and the column on the right, as shown in Fig. 12-7b. Because of this difference in end conditions, each weld must receive individual consideration. As a means of identification, the different welds will be referred to as *A*, *B*, *C*, and *D*, as indicated in Fig. 12-7b.

The negative bending moment in the beam at the supports is 1150 kip-in., and E 70 XX electrodes will be used. Design the welds.

Solution: (1) The resultant tensile and compressive stresses in the flanges constitute a *mechanical couple*. A mechanical couple consists of two equal parallel forces, opposite in direction, not having the same line of action. The moment of a couple is the magnitude of one of the forces multiplied by the normal distance between their lines of action. Therefore, if the negative bending moment is 1150 kip-in. and the beam depth is 12 in., the resultant tensile and compressive stresses in the flanges at the supports are each 1150 ÷ 12 = 96 kips.

(2) Referring to Table 1-1, we find that the W 12 × 40 has a flange width of 8 in. and a flange thickness of 0.516 in. First, suppose we run a ⅜-in. fillet weld across the upper flange for weld A. This weld has an allowable working stress of 5.6 kips per lin in. (Table 12-1) and, as the flange is 8 in. wide, its resistance is 8 × 5.6 = 44.8 kips. If ⅜-in. fillet welds are to be used at this joint, the total length of weld must be 96 ÷ 5.6 = 17.2 in. But the flange is only 8 in. wide, and welds on both the upper and under surfaces of the flange, as shown in Fig. 12-7c, will not afford a sufficient length. A solution for this condition would be to investigate larger welds, but the weld at the underside of the flange requires overhead welding, which should be avoided whenever possible. It should be remembered that this connection is in tension and that the strength of a fillet weld is determined by shear at the throat of the weld.

(3) Another method is to weld a ¼-in. plate 7 in. wide to the upper face of the beam flange (Fig. 12-7d and e). The flange and plate are bevel-cut, and a butt joint is used. The resistance to tension of the flange weld is 0.516 × 8 × 22 = 91 kips, and for the auxiliary plate it is 0.25 × 7 × 22 = 38.5 kips. This gives a total resistance of 91 + 38.5 = 130 kips, which is in excess of the required 96 kips. The weld shown in Fig. 12-7d and e is therefore acceptable. These computations serve to illustrate a procedure that is sometimes used, although in this instance the plate width of 7 in. could be reduced in order to provide less excess capacity.

(4) The same weld is impracticable for joint B at the opposite end of the beam because of the open space adjacent to the column. At this joint, a ⅜-in. plate will be used and a butt joint employed to connect it to the column (Fig. 12-7f and g). The required length of weld to give a resistance of 96 kips is 96 ÷ (0.375 × 22) = 11.6 in. This indicates that the plate must have a greater width than the beam flange at the face of the column, so it will have a shape as

shown in Fig. 12-7g. To determine the length of fillet weld required to secure the plate to the beam flange, we will assume $\frac{5}{16}$-in. welds having an allowable working strength of 4.6 kips per lin in. (Table 12-1). The required design length is 96 ÷ 4.6 = 20.8 in., half of which will be placed on each edge of the plate.

(5) At joint C, the lower flange of the beam exerts compressive stress directly against the column face. A $\frac{1}{4}$-in. fillet weld is placed on the upper surface of the lower flange to hold the beam in position (Fig. 12-7h).

(6) For joint D, a butt joint is used between the lower flange and the column, as shown in Fig. 12-7i. Its resistance to compression is 0.516 × 8 × 22 = 91 kips. This is less than the 96 kips required, so the detail shown in Fig. 12-7i will have to be modified to provide additional capacity. One method of doing this is to make the seat angle a "working" angle, adding fillet welds along the edges of the beam flange as indicated in the top sketch of Fig. 12-9b.

Beams welded to columns to produce continuous action exert compressive and tensile forces on the supports to such a degree that reinforcement in the columns is necessary. This is accomplished by welding plates between the flanges of the columns, as shown in Fig. 12-7j. The welds in the upper and lower plates resist tensile and compressive stresses, respectively.

12-7 Plug and Slot Welds

One method of connecting two overlapping plates uses welds in holes made in one of the two plates. See Fig. 12-8. Generally, plug and slot welds are welds in which the entire area of the hole or slot receives weld metal. A somewhat similar weld consists of a fillet weld at the circumference of a hole, as shown in Fig. 12-8c. Plug or slot welds are used to transmit shear in a lap joint. The maximum and minimum diameters of plug and slot welds and the maximum length of slot welds are shown in Fig. 12-8. If the plate containing the hole is not more than $\frac{5}{8}$ in. thick, the hole should be filled with weld metal. If the plate is more than $\frac{5}{8}$ in. thick, the weld metal should be at least one-half the thickness of the material but not less than $\frac{5}{8}$ in.

The stress in a plug or slot weld is considered to be shear on the

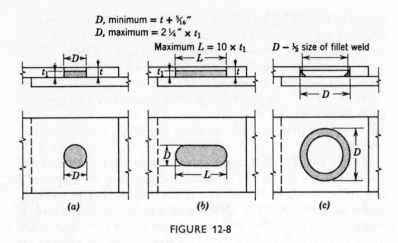

FIGURE 12-8

area of the weld at the plane of contact of the two plates being connected. The allowable unit shearing stress, when E 70 XX electrodes are used, is 21 ksi.

Example. A plug weld consists of weld metal in a $1\frac{1}{4}$-in. diameter hole in a plate $\frac{1}{2}$ in. thick, as indicated in Fig. 12-8a. Compute the load this weld will transmit.

Solution: The effective weld area is that of the $1\frac{1}{4}$-in. diameter circle, or $D^2 \times 0.7854 = 1.25^2 \times 0.7854 = 1.227$ sq in. The allowable shearing stress of the weld metal is 21 ksi; therefore, $1.227 \times 21 = 25.8$ kips, the load the plug weld will transmit.

Problem 12-7-A. A plug weld consists of a $1\frac{1}{8}$-in. diameter hole filled with weld metal deposited from an E 70 XX electrode. What load will the weld transmit?

Problem 12-7-B.* The hole for a slot weld is 1 in. wide and 6 in. long with the ends semicircular, as shown in Fig. 12-8b. Compute the load the slot weld will transmit if it is made with E 70 XX electrodes.

12-8 Miscellaneous Welded Connections

Part 4 of the AISC Manual contains a series of tables pertaining to design of welded connections. The tables cover free-ended as well as moment-resisting connections for use in Type 2 and Type 1 construction, respectively. In addition, suggested framing and column splicing details are presented.

A few miscellaneous connections are shown in Fig. 12-9. As an aid in erection, certain parts are welded together in the shop before being sent to the site. Connection angles may be shop-welded to beams and the angles field-welded to girders or columns. The beam connection indicated in Fig. 12-9a shows a beam supported on a seat that has been shop-welded to the column. A small connection plate is shop-welded to the lower flange of the beam, and the plate is bolted to the beam seat. After the beams have been erected and the frame plumbed, the beams are field-welded to the seat angles. This type of beam connection provides no degree of continuity, there being no restraint and hence no bending moment developed at the support.

Continuous action results from the connections shown in Fig. 12-9b and c. Auxiliary plates are used to make the connection at the upper flange.

Beam seats shop-welded to columns are shown in Fig. 12-9d, e, and f. A short length of angle welded to the column with no stiffeners is shown in Fig. 12-9d. Stiffeners (triangular-shaped plates) are welded to the legs of the angles shown in Fig. 12-9f and add materially to the strength of the seat. Another method of forming a beam seat is to use a T-section or half an I-beam cut as indicated in Fig. 12-9e.

Various types of column splices are shown in Fig. 12-9g, h, i, and j. The auxiliary plates and angles are shop-welded to the columns and provide for bolted connections in the field before making the permanent welds.

Figure 12-9k shows a type of welded connection used in trusses where the lower chord is composed of a structural T-section or a split I-beam. The angle sections of the truss web members are welded directly to the stem of the tee. Connections in trusses are considered in Chapter 13.

Figure 12-9l and m shows welded connections for columns and base plates. The angles are shop-welded to the columns and the columns are field-welded to the base plates.

Additional connection details are shown in Fig. 12-10. The detail of Fig. 12-10a illustrates an arrangement for framing a beam to a

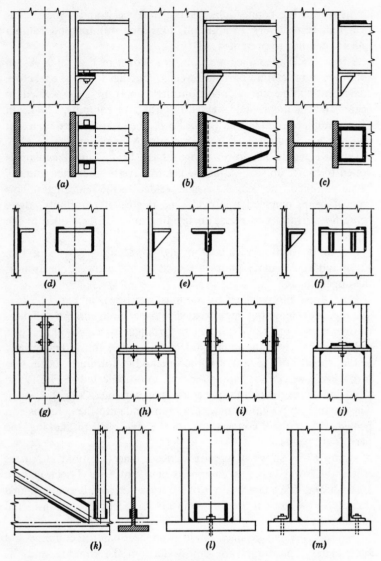

FIGURE 12-9

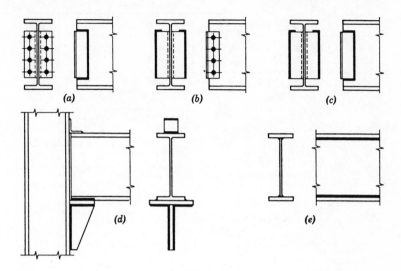

FIGURE 12-10

girder in which welds are substituted for bolts or rivets. In this figure, welds replace the fasteners in securing the connection angles to the web of the beam. In Fig. 12-10*b*, welds replace rivets or bolts in securing the connection angles to the web of the girder (not shown in the illustration).

Figure 12-10*c* gives the positions of welds for connection angles welded to both the beam and the girder. A welded connection for a stiffened seated beam connection to a column is shown in Fig. 12-10*d*. Figure 12-10*e* shows the simplicity of welding in connecting the upper and lower flanges of a plate girder to the web plate. (Compare Fig. 8-16.)

12-9 Symbols for Welds

In making detail drawings of welded connections of structural elements, standard symbols are used. In addition to the type of weld, other information to be conveyed includes size, exact location, finishes, etc. Figure 12-11, reproduced from the AISC Manual, gives the standard symbols for welded joints. It will be noted that

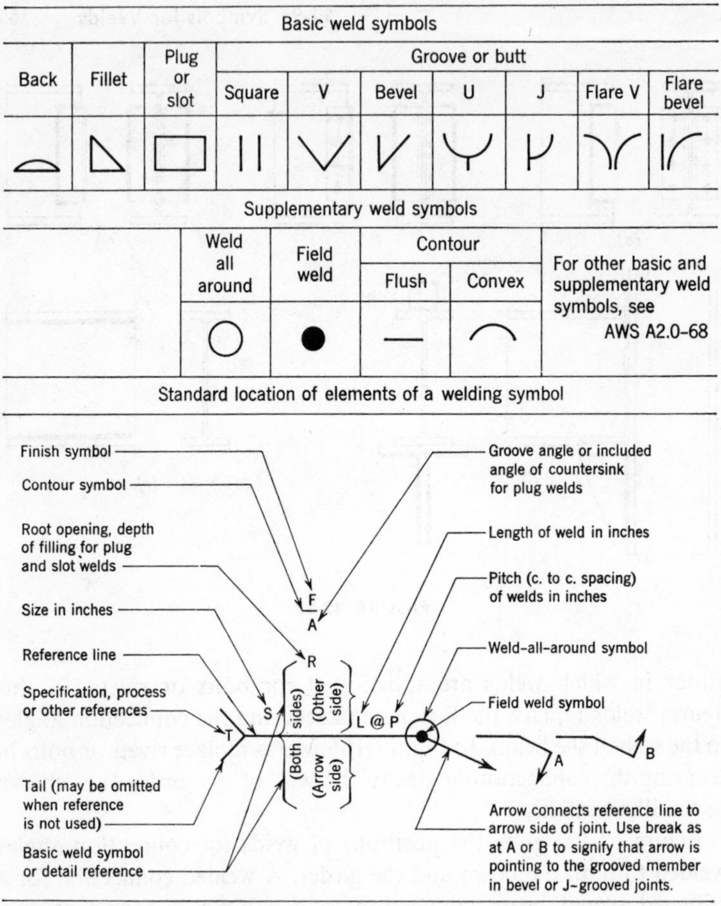

Basic weld symbols

Back	Fillet	Plug or slot	Groove or butt						
			Square	V	Bevel	U	J	Flare V	Flare bevel
⌒	△	▢	‖	∨	�V	Y	⊬	⋎	⎷

Supplementary weld symbols

	Weld all around	Field weld	Contour		For other basic and supplementary weld symbols, see AWS A2.0-68
			Flush	Convex	
	◯	●	—	⌒	

Standard location of elements of a welding symbol

Finish symbol

Contour symbol

Root opening, depth of filling for plug and slot welds

Size in inches

Reference line

Specification, process or other references

Tail (may be omitted when reference is not used)

Basic weld symbol or detail reference

Groove angle or included angle of countersink for plug welds

Length of weld in inches

Pitch (c. to c. spacing) of welds in inches

Weld–all–around symbol

Field weld symbol

Arrow connects reference line to arrow side of joint. Use break as at A or B to signify that arrow is pointing to the grooved member in bevel or J-grooved joints.

Note:

Size, weld symbol, length of weld and spacing must read in that order from left to right along the reference line. Neither orientation of reference line nor location of the arrow alter this rule.

The perpendicular leg of △, ∨, ⊬, ⎷, weld symbols must be at left.

Arrow and Other Side welds are of the same size unless otherwise shown.

Symbols apply between abrupt changes in direction of welding unless governed by the "all around" symbol or otherwise dimensioned.

These symbols do not explicitly provide for the case that frequently occurs in structural work, where duplicate material (such as stiffeners) occurs on the far side of a web or gusset plate. The fabricating industry has adopted this convention; that when the billing of the detail material discloses the identity of far side with near side, the welding shown for the near side shall also be duplicated on the far side.

FIGURE 12-11 Standard symbols for welded joints. Taken from the 2nd Edition of *Structural Steel Detailing*. Courtesy American Institute of Steel Construction.

the symbol for a fillet weld is a triangle; this is drawn below the horizontal line if the weld is on the near side, above if it is on the far side; two triangles, one above and one below, are drawn for welds on both sides of the joint. The size of the weld is placed to the left of the vertical line of the triangle and the length to the right side of the hypotenuse. Figure 13-8b shows how this symbol is used to indicate the welding required in a lower-chord joint of a truss.

13

Roof
Trusses

IIIIIIIIIIIIIIIIIIIIIIIIIIIIIIIIIIIIIIIIIIIIIIII

13-1 General

A truss is a framed structure, usually supported only at its ends, with a system of members so arranged and secured to each other that the stresses transmitted from one member to another are longitudinal only—that is, the members are in either compression or tension. Basically, a truss is composed of a system of triangles, since a triangle is the only polygon whose shape is incapable of being changed without changing the length of one or more of its sides.

That portion of a roof that occurs between two adjacent trusses is called a *bay*, the spacing of the trusses on centers being the width of a bay. A *purlin* is the beam spanning from truss to truss that transfers to the trusses the loads due to the weight of the roof deck and roofing, snow, and wind. The portion of a truss that occurs between two adjacent joints of the upper chord is called a *panel*. The load brought to an upper-chord joint or *panel point* is, then, the roof load in pounds per square foot multiplied by the panel length times the bay width. Figure 13-1 identifies the several elements of a typical roof truss.

The height or *rise* of the roof divided by the span is called the *pitch;* the rise divided by half the span is the *slope*. These two terms

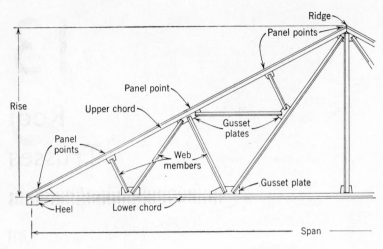

FIGURE 13-1   Parts of a typical roof truss.

are often used interchangeably and thus incorrectly, especially by workmen in the field. A clearer way of expressing slope is to give the amount of rise per foot of horizontal run. A roof that rises 6 inches in a horizontal distance of 12 inches has a slope of "6 in 12." Reference to Table 13-1 should clarify this nomenclature.

TABLE 13-1.   Roof Pitches and Slopes

Pitch	⅛	⅙	⅕	¼	1/3.46	⅓	½
Degrees..	14° 3′	18° 26′	21° 48′	26° 34′	30° 0′	33° 40′	45° 0′
Slope....	3 in 12	4 in 12	4.8 in 12	6 in 12	6.92 in 12	8 in 12	12 in 12

When the rise is not greater than about 2 inches per horizontal foot, the roof is said to be flat. Very often the slope is determined by aesthetic considerations, or it may be controlled by the type of roofing material. For commercial buildings, the slope is usually determined by economic considerations. It is generally considered that a rise of about 6 inches to the foot is probably the most economical for average spans. Roofs with steeper slopes require the use of more roofing material, while slopes less steep produce increased

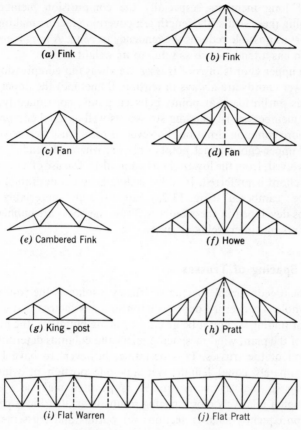

FIGURE 13-2 Types of roof trusses.

stresses in the truss members. Figure 13-2 shows some of the more common types of roof trusses.

The type of truss to use depends on a number of factors such as the length of span, the roofing material, and the magnitude of the loads. The span of the truss and the material of which the roof deck is constructed determine the number of panels, which in turn influences the choice of truss type. Some types of trusses are more economical than others; and in this respect, experience and practical considerations constitute the best guide. It is always advisable to avoid the use of long members, especially the compression members. In designing these struts, the length is a governing factor; making them as short as possible reduces the tendency to bend. A long member in tension has a tendency to sag due to its weight.

The upper chords of roof trusses are always in compression, and the lower chords are always in tension. Sometimes the upper chord receives purlin loads at points between joints; consequently, those chord members resist bending stresses as well as axial compression. The more common situation, however, is for a truss to receive loads at the upper-chord panel points only or, where suspended ceilings are involved, from the lower-chord joints also. The use of a horizontal lower chord is preferred. If added ceiling height is desirable, a truss may be "cambered" (Fig. 13-2e), but this adds appreciably to the stresses developed in the members. Truss members are connected at the joints by either welds or rivets.

13-2 Spacing of Trusses

The most economical spacing of trusses insofar as the roof system as a whole is concerned depends upon so many factors that a simple rule for finding it cannot be given. For many buildings, the arrangement of the plan, window spacing, piers and columns determines the location of the trusses. It is desirable, however, to have bays of approximately equal length over any one portion or wing of a building so that as many trusses as possible may be identical. Although a relatively close spacing results in smaller loads per truss and consequently smaller sections for purlins and truss members, a greater number of trusses increases the expense involved in fabrication. With loadings approximating 30 psf and for spans up to about 50 ft, a spacing of 15 to 18 ft is generally satisfactory.

13-3 Loads on Roof Trusses

The first step in the design of a roof truss consists of computing the loads the truss will be required to support. These comprise both dead and live loads. The former includes the weight of all construction materials supported by the truss, and the latter includes loads resulting from snow and wind and, on flat roofs, occupancy loads and an allowance for the possible ponding of water due to impaired drainage.

The following items constitute the materials to be considered in computing the dead loads: roof covering and roof deck, purlins and sway bracing, ceiling and any suspended loads, and the weight of the truss itself. Obviously, all these loads cannot be exactly determined before the truss is designed, but all may be checked later to see whether a sufficient allowance has been made. The dead loads are downward vertical forces, and hence the end reactions of the truss are also vertical with respect to these loads. Table 9-1 gives the weights of certain roofing materials, and Table 13-2 provides

TABLE 13-2. Approximate Weight of Steel Trusses in Pounds per Square Foot of Roof Surface

Span	Slope of roof			
Feet	45°	30°	25°	Flat
Up to 40	5	6	7	8
40–50	6	7	7	8
50–60	7	8	9	10
60–70	7	8	9	10
70–80	8	9	10	11

estimated weights of steel trusses for various spans and pitches. With respect to the latter, one procedure is to establish an estimate in pounds per square foot of roof surface and consider this load as acting at the panel points of the upper chord. A more exact method would be to apportion a part of such loads to the panel points of the lower chord, but this is customary only in trusses having exceptionally long spans. After the truss has been designed, its actual weight may be computed and compared with the estimated weight. If the two are not in reasonable agreement, another weight should be assumed and the truss redesigned. Except for unusual cases, this step is seldom necessary.

13-4 Snow Load

The weight allowance for snow load depends primarily upon the geographical location of the structure and the roof slope. Freshly fallen snow may weigh as much as 10 lb per cu ft, and accumulations of wet or packed snow may exceed this value. The amount of snow retained on a roof over a given period depends upon the type of roofing as well as the slope. For example, snow slides off a metal or slate roof more readily than from a wood shingle surface; also, the amount of insulation in the roof construction will influence the period of retention. The local building code should always be consulted with respect to mandatory snow load allowances. Where there is no building code or recognized local custom to serve as a guide, Table 13-3 will be helpful in establishing a proper snow load

TABLE 13-3. Snow Loads for Roof Trusses in Pounds per Square Foot of Roof Surface

Locality	Slope of roof				
	45°	30°	25°	20°	Flat
Northwestern and New England states	15	20	30	35	40
Western and Central states	10	15	25	30	35
Pacific and Southern states	0	5	10	10	10

allowance. It should be noted that the values given in Table 13-3 are for pounds per square foot of *roof surface*.

Some building codes require that, for a roof of any configuration, whether flat, curved, or pitched, an allowance of 20 psf of horizontal projection be included in addition to the dead load and wind load. This minimum load is intended to provide for loads incidental to construction and repair as well as for minor snow loads that may occur in areas not normally subject to snowfall. This point should be investigated when using values from Table 13-3.

Since the snow load acts vertically, some designers add this load to the dead load for the purpose of drawing a vertical load stress diagram (Art. 13-6). Frequently, however, it is desirable to know the stresses in members due to snow loads alone. In this case, a separate snow load stress diagram may be drawn, or the stresses may be computed from the stresses due to dead load, as they are directly proportional.

13-5 Wind Load

Except under hurricane conditions, the wind pressure directed at right angles to the side of a building of moderate height probably never exceeds 40 psf. A great deal of research has been carried on in recent years related to the dynamic effect of wind on buildings. The use of lighter weight materials for curtain walls and roofs has made it necessary to take into account the negative pressures developed by wind force in order to design against building surfaces being blown outward as well as inward. The American Standard Building Code Requirements for Minimum Design Loads in Buildings and Other Structures (ANSI A58.1-1955)[1] calls for roofs with slopes greater that 30 degrees to be designed to withstand inward pressures, acting normal to the windward surface only, equal to those specified for the height zone in which the roof is located. The Code also provides that roofs shall be designed to withstand pressures acting outward normal to the surface, equal to $1\frac{1}{4}$ times those specified for the corresponding height zone. This latter provision is, of course, important in designing the fastenings that secure the roof deck and covering to the purlins.

An extended treatment of wind pressures and wind effects is beyond the scope of this text. The discussion here will be based on the premise that the effective wind pressure acts normal (perpendicular) to the roof surface and varies with the slope as indicated in Table 13-4. The values in the table are based on a wind pressure

[1] American National Standards Institute, Inc. Excerpts from this document are contained in Part 5 of the AISC *Manual*, 7th Edition.

TABLE 13-4. Wind Pressure on Roof
Surfaces

Slope of roof	Normal pressure
Degrees	Pounds per square foot
15	14.0
20	18.0
25	22.0
30	26.0
35	30.0
40	33.0
45	36.0
50	38.0

of 40 psf on a vertical surface. Since the wind load is an inclined
force on the truss, the truss reactions will not be vertical for wind
loading. This point is discussed further in Art. 13-7.

13-6 Stress Diagrams

After the panel loads on a truss have been established, the next step
in design is to determine the stresses in the members. This may be
accomplished by analytic or graphic methods, but only the graphic
method of *stress diagram* will be considered here.

Figure 13-3a shows panel loads on a truss for both vertical and
wind loading. Wind-load stress diagrams will be discussed in Art.
13-7; our concern here will be with construction of a stress diagram
for the vertical loads only. Since each panel load is 4 kips, the half-
panel loads at the ends are each 2 kips, making a total vertical load
of $4 + 4 + 4 + 2 + 2 = 16$ kips. Because the truss is symmetrical,
each end reaction is $16 \div 2 = 8$ kips.

The panel loads and reactions being known, *the first step in
constructing a stress diagram is to draw a force polygon of the external
forces.* These forces are *AB, BC, CD, DE, EF, FG,* and *GA,* and
the magnitudes of all are known. Therefore, at a suitable scale,
draw *ab* (Fig. 13-b), a downward force equal to 2 kips. The next
external force is *BC,* and from point *b,* just determined, draw *bc*

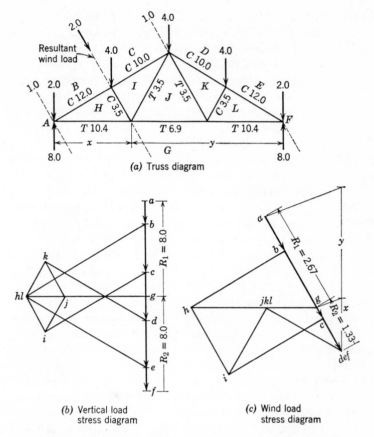

FIGURE 13-3   Truss and stress diagrams (loads and stresses in kips).

downward, equivalent to 4 kips. Continue with *CD*, *DE*, and *EF*. This completes the downward forces. The line just drawn is called the load line, the length being equivalent to 16 kips. The next external force is *FG*, an upward force of 8 kips. This determines the position of point *g*; and *GA*, the remaining external force, completes the polygon of the external forces. Because the loads and reactions are vertical, the force polygon of the external forces is a vertical line.

In conjunction with the polygon just drawn, draw a force polygon for the forces *AB*, *BH*, *HG*, and *GA* about the point *ABHG*. From *b*, draw a line parallel to *BH*; and from *g*, draw a line parallel to *HG*. Their intersection determines point *h*. Next consider the members about the joint *BCIH*. From *c*, draw a line parallel to *CI*; and through *h*, draw a line parallel to *IH*, their intersection determining point *i*. The next joint is *HIJG*. Through *i*, draw a line parallel to *IJ*; and through *g*, draw a line parallel to *JG*, thus establishing point *j*. In the same manner, take the joints *CDKJI* and *DELK*. This completes the stress diagram. See Fig. 13-3*b*. The magnitudes of the stresses in the members are found by scaling the lengths of the lines in the stress diagram just completed.

It is now necessary to determine the character of the stress in each member—that is, whether the member is in tension or compression. A minus sign (−) usually denotes compression, and a plus sign (+) denotes tension. However, in some books, these designations are reversed. In order to avoid confusion, the system employed in this text uses the symbol (*C*) for compressive stress and (*T*) for tensile stress. The important point to remember is that a member in compression tends to be made shorter and resists this shortening by *pushing against* the joints at its ends. A tension member, on the other hand, tends to become longer and resists the lengthening by *pulling away* from its end joints.

Consider first the upper chord member *BH* at the heel joint *ABHG* in Fig. 13-3*a*. Because the letters designating the members were read in a clockwise direction around the joint, note the sequence: *B* first, then *H*. In the stress diagram, we find that *bh* reads downward to the left or *toward* the joint; hence, member *BH* is in compression.

Using the heel joint again to find the nature of the stress in the lower chord member entering the joint, we read its designation clockwise as *HG*. In the stress diagram, we find that *hg* reads from left to right or *away from* the joint; therefore, member *HG* is in tension.[2] The character of the stress in each of the remaining members is found in a similar manner.

---

[2] The reader should satisfy himself that the same character of stress will be obtained for these two members by following the clockwise reading procedure at the upper chord joint *BCIH* and the lower chord joint *HIJG*, respectively.

It should be noted that the length of a member in the truss diagram (Fig. 13-3a) bears no direct relation to the magnitude of its stress. The magnitude of the stress is determined by the length of the line in the stress diagram corresponding to the truss member. For vertical loads, the character and magnitude of the stresses in all members are shown on the truss diagram.

13-7 Wind Load Stress Diagram

The stresses in members of a roof truss due to wind loads are found by constructing a stress diagram as previously described. It is assumed that the wind exerts a pressure perpendicular to the roof surface. In the truss diagram (Fig. 13-3a), the wind, indicated by dotted lines, is shown coming from the left, the total load being $1 + 2 + 1 = 4$ kips. To draw a stress diagram for the wind loads, proceed as before. Construct the force polygon of the external forces, namely AB, BC, C–DEF, DEF–G, and GA, the latter two being the wind-load reactions. It will be noted that, because the wind comes from the left, there are no forces DE and EF; consequently, the letters D, E, and F represent a single point in the stress diagram. Draw ab, bc, and c–def. See Fig. 13-3c. It may be assumed that the reactions due to wind loads are parallel to the direction of the wind; and because the wind comes from the left, the left-hand reaction will have a greater magnitude than that on the right for this particular truss. For the purpose of finding the magnitudes of the reactions, we may consider that all the wind loads are concentrated in a single line of action at BC. The resultant wind load continued divides the lower chord into two parts, x and y, and the magnitudes of the reactions are to each other as the division x is to y. The line a-def, representing the total wind load, is therefore divided in the same proportion as the two divisions of the lower chord, x and y. To accomplish this, erect a line from the point def in a length equivalent to the length of the lower chord and divide it into sections x and y. From the upper extremity of the line just drawn, draw a line to point a and a parallel line from the point separating x and y to the load line. This determines point g and, consequently, the reactions R_1 and R_2. The force polygon of the external forces is now completed, and the stress diagram is then drawn as previously described. It will be found that the letters j, k, and l fall at one point, indicating that there are no stresses in JK and KL when the wind comes from the left. For design purposes, however, the stresses in JK and KL are taken the same as for JI and IH, respectively, since the direction of the wind may be reversed.

13-8 Combinations of Loading

It is improbable that a maximum snow load and maximum wind load will occur simultaneously on a roof. The combinations of loads most likely to occur and commonly used by many designers are as follow:

1. Dead load and maximum snow load
2. Dead load, maximum wind, and minimum snow
3. Dead load, minimum wind, and maximum snow

To determine accurately the greatest stresses that the various truss members will be required to resist, it is necessary to determine separately the stresses due to dead, snow, and wind loads. These are then tabulated for each member, together with a tabulation for each of the combinations listed above. The combination giving the greatest stress in a member is the one used for design. It is customary to consider one half the maximum snow and wind loads as the minimum values.

13-9 Equivalent Vertical Loading

A convenient and practical method of computing stresses in truss members is to use an equivalent vertical load for the combined snow and wind loads. For the usual simple roof trusses of moderate span supported on masonry walls, the maximum stresses in the members determined from the combinations stated in Art. 13-8 are substantially the same as for a uniform vertical load acting over the entire roof surface. The method of equivalent vertical loading requires only one stress diagram to be drawn and will answer for any probable combination of dead, wind, and snow loads when the proper equivalent vertical live load is chosen. Many building codes specify a minimum vertical live load that must be used for buildings within their jurisdiction; but in the absence of mandatory provisions, selection of the proper equivalent load depends on the judgment and experience of the designer. Table 13-5 may be used as a guide in this

TABLE 13-5.   Equivalent Vertical Loads for Combined Snow and Wind Loads in Pounds per Square Foot of Roof Surface

Locality	Slope of roof				
	45°	30°	25°	20°	Flat
Northwestern and New England states	28	25	24	35	40
Western and Central states	28	25	24	30	35
Southern and Pacific states	28	25	24	22	20

connection, but local knowledge of snowfall and wind velocity may make it prudent to increase these values in individual cases.

13-10   Stresses Found by Coefficients

The stress diagram provides a relatively simple analysis of the stresses in truss members. It can be used for symmetrical or unsymmetrical loading for any type of truss, and it is self-checking. In practice, however, there are numerous trusses of moderate span with symmetrical vertical loads only on the upper chord. Table 13-6 shows six commonly used trusses and lists coefficients for determining the stresses in the members. Four different pitches are given for each truss: $\frac{1}{3}$, 30°, $\frac{1}{4}$, and $\frac{1}{5}$. As shown in Table 13-1, a slope of 30° corresponds to a pitch of $\frac{1}{3.46}$, the pitch being the ratio of the height to the span of the truss. For the trusses shown in the diagrams, the heavy lines indicate compression members, and the lighter lines indicate tension members.

Table 13-6 affords a ready means of establishing the stresses in the trusses shown. It avoids the necessity of constructing stress diagrams. The coefficients given in the tabulations have been determined analytically, and the stress in any truss member can be

TABLE 13-6. Coefficients for Determining Stresses in Simple Trusses*

Simple Fink Truss

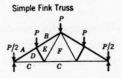

Member	Character of Stress	Pitch = height over span ⅓	30°	¼	⅕
AD	Compression	2.70	3.00	3.35	4.04
BE	Compression	2.15	2.50	2.91	3.67
DC	Tension	2.25	2.60	3.00	3.75
FC	Tension	1.50	1.73	2.00	2.50
DE	Compression	0.83	0.87	0.90	0.93
EF	Tension	0.75	0.87	1.00	1.25

King Post Truss

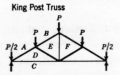

Member	Character of Stress	Pitch = height over span ⅓	30°	¼	⅕
AD	Compression	2.70	3.00	3.35	4.04
BE	Compression	1.80	2.00	2.24	2.69
DC	Tension	2.25	2.60	3.00	3.75
DE	Compression	0.90	1.00	1.12	1.35
EF	Tension	1.00	1.00	1.00	1.00

Howe Truss

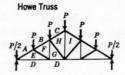

Member	Character of Stress	Pitch = height over span ⅓	30°	¼	⅕
AE	Compression	4.51	5.00	5.59	6.73
BF	Compression	3.61	4.00	4.50	5.39
CH	Compression	2.70	3.00	3.30	4.04
ED	Tension	3.75	4.33	5.00	6.25
GD	Tension	3.00	3.46	4.00	5.00
EF	Compression	0.90	1.00	1.10	1.35
FG	Tension	0.50	0.50	0.50	0.50
GH	Compression	1.25	1.32	1.40	1.60
HI	Tension	2.00	2.00	2.00	2.00

Pratt Truss

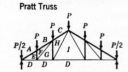

Member	Character of Stress	Pitch = height over span ⅓	30°	¼	⅕
AE	Compression	4.51	5.00	5.59	6.73
BF	Compression	4.51	5.00	5.59	6.73
CH	Compression	3.61	4.00	4.47	5.39
ED	Tension	3.75	4.33	5.00	6.25
GD	Tension	3.00	3.46	4.00	5.00
ID	Tension	2.25	2.60	3.00	3.75
EF	Compression	1.00	1.00	1.00	1.00
FG	Tension	1.25	1.32	1.41	1.60
GH	Compression	1.50	1.50	1.50	1.50
HI	Tension	1.68	1.73	1.80	1.95

Fan Truss

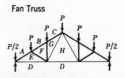

Member	Character of Stress	Pitch = height over span ⅓	30°	¼	⅕
AE	Compression	4.51	5.00	5.59	6.72
BF	Compression	3.54	4.00	4.55	5.57
CG	Compression	3.40	4.00	4.71	5.98
ED	Tension	3.75	4.33	5.00	6.25
HD	Tension	2.25	2.60	3.00	3.75
EF	Compression	0.93	1.00	1.08	1.21
FG	Compression	0.93	1.00	1.08	1.21
GH	Tension	1.50	1.73	2.00	2.50

Fink Truss

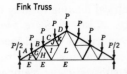

Member	Character of Stress	Pitch = height over span ⅓	30°	¼	⅕
AF	Compression	6.31	7.00	7.83	9.42
BG	Compression	5.76	6.50	7.38	9.05
CJ	Compression	5.20	6.00	6.93	8.68
DK	Compression	4.65	5.50	6.48	8.31
FE	Tension	5.25	6.06	7.00	8.75
HE	Tension	4.50	5.20	6.00	7.50
LE	Tension	3.00	3.46	4.00	5.00
FG	Compression	0.83	0.87	0.89	0.93
GH	Tension	0.75	0.87	1.00	1.25
HI	Compression	1.66	1.73	1.79	1.86
IJ	Tension	0.75	0.87	1.00	1.25
JK	Compression	0.83	0.87	0.89	0.93
KL	Tension	2.25	2.60	3.00	3.75
IL	Tension	1.50	1.73	2.00	2.50

* Reproduced from *Simplified Design of Roof Trusses* by Harry Parker, published by John Wiley and Sons, N. Y.

obtained by multiplying the panel load P by the appropriate coefficient found in the column giving the pitch of the roof.

To show how readily stresses may be found by use of this table, consider the following illustration: a Howe 6-panel truss with a $\frac{1}{4}$ pitch is to be constructed, and P, the panel loads, are each 7400 lb.

We consult Table 13-6 for the magnitude of the stress in the member of the upper chord adjacent to the support (the heel joint) and also for the character of the stress in this member. By referring to the Howe truss, we identify the member in question as AE. For a $\frac{1}{4}$ pitch, the coefficient for this member is 5.59. Therefore, $5.59 \times 7400 = 41{,}360$ lb, the stress in this member of the upper chord. Stresses in the other members of the truss are found similarly by multiplying the panel load by the appropriate coefficients in the table. As for the character of stress in member AE, we see in the truss diagram that AE is a heavy line; hence, the member is in compression.

In all simple triangular trusses, the upper-chord members resist compressive stresses, and the stresses in the lower chord are in tension. There are no rules for determining the character of the stresses in the web members. These are readily determined if a stress diagram has been drawn, as demonstrated in Art. 13-6.

13-11   Design of a Roof Truss

The various steps in the design of a steel roof truss are illustrated by subsequent articles of this chapter, which relate to the solution of the example stated below.

Example.   Design a steel, eight-panel Fink truss having a span of 48 ft and a height of 12 ft; this constitutes a $\frac{1}{4}$ pitch. The trusses are spaced 17 ft on centers, and the roof construction consists of precast lightweight concrete slabs, weighing 32 lb per sq ft, that span between the purlins. The roofing will be $\frac{1}{4}$-in. slate. The truss, which is for a building located in the New England States, will be fabricated from A36 steel and will consist primarily of double-angle members connected to $\frac{3}{8}$-in. gusset plates. The 1969 AISC Specification controls the design.

Figure 13-4a represents the truss diagram drawn to scale. For this

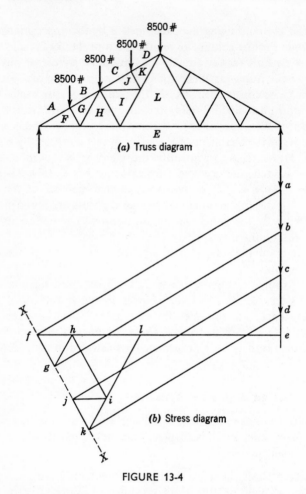

(a) Truss diagram

(b) Stress diagram

FIGURE 13-4

span and rise, the distance between panel points on the upper chord is approximately 6.7 ft.

13-12   Computation of Truss Loads

The area of roof supported by one purlin is equal to the bay length (truss spacing) times the panel length or $17 \times 6.7 = 114$ sq ft. The load on a purlin is made up of the weight of the roofing and deck construction plus the equivalent vertical load for combined snow and wind. Table 9-1 gives the weight of ¼-in. roofing slate as 9.5 psf and, by data, the precast deck slabs weigh 32 psf. To establish the vertical equivalent snow and wind loading, we note from Table 13-1 that a ¼ pitch corresponds to a slope of approximately 26½°. Referring to Table 13-5, the value for a 25° slope is 24 psf, and the value for a 30° slope is 25 psf. Use, say, 24.5 psf. The load brought to the purlin is then

$$
\begin{aligned}
\text{Slate roofing} &= 9.5 \\
\text{Roof deck} &= 32.0 \\
\text{Equivalent snow and wind} &= 24.5 \\
\hline
\text{Total} &= \overline{66.0} \text{ psf}
\end{aligned}
$$

The vertical panel load, not including the weight of the purlin or truss, is $114 \times 66 = 7520$ lb. This is the load used to design the purlins. When the weight of the purlins is known, an allowance for weight of the truss and bracing will be established and the total load brought to each panel point of the truss computed.

13-13 Purlin Design

When purlins are attached to the upper chord of roof trusses, the X–X axis of the purlin is usually parallel to the chord member. If the load on the purlin is vertical and the purlin is free to bend in any direction, bending will not take place about either the X–X or Y–Y axis. This is called unsymmetrical bending (Fig. 13-5a). Sometimes

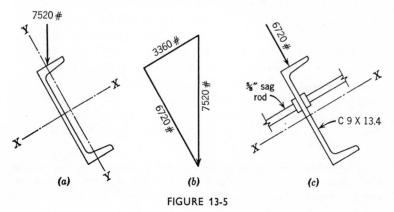

FIGURE 13-5

the roof deck construction is of a type that provides restraint to bending in a direction parallel to the roof surface, but the precast slabs used in this example cannot be depended upon to act in this manner. The installation of *sag rods*, as shown in Fig. 13-5c, will prevent bending parallel to the roof surface; and the purlins may be designed for the normal component of the vertical load only, with bending about the X–X axis. Sag rods are usually ⅝ or ¾ in. in diameter and are placed in two rows between adjacent trusses at the third points of the purlin span. They run from the purlin nearest the heel joint on one side of the truss, up over the ridge, and down the other side. Since the stress in these rods is greatest at the ridge, it is frequently necessary to increase the size of the ridge purlin in order to provide for the concentrated load due to the vertical component of the sag-rod stresses. Sag rods may also be used with roof deck construction that is rigidly attached to purlins to provide lateral support during erection. They are seldom necessary on roofs having slopes less than 3 in 12.

In Fig. 13-5b, the vertical panel load supported by one purlin (7520 lb) is drawn to a suitable scale. The components of this force, one parallel and one normal to the roof slope, are drawn and are found to measure 3360 and 6720 lb, respectively. The force parallel to the roof slope (3360 lb) is resisted by the sag rods; the other force (6720 lb) produces bending about the X–X axis of the purlin.

The force of 6720 lb constitutes a uniformly distributed load over the 17-ft span of the purlin. The fastenings of the precast roof deck will be assumed to furnish lateral support to the top flange of the channel purlins;[3] therefore, we can use 22 ksi as the bending stress, and the purlin may be designed as a simple beam. Referring to Table 8-3, we find that an 8-in channel weighing 11.5 lb per lin ft will support 7000 lb on a 17-ft span but with a total deflection exceeding ⅟₃₆₀ of the span. Although deflection is not usually a factor in purlin design, it is often prudent to select a member in accordance with the approximate rule of thumb that states that the depth of a beam in inches should not be less that half the span in

---

[3] The lighter weight Miscellaneous sections (M Shapes) are also used for purlins. Many of these sections have wider flanges than channels and lend themselves readily to the attachment of precast roof deck systems.

feet. Table 8-3 shows that a C 9 × 13.4 is the lightest weight section that meets this condition.

Because the purlins weigh 13.4 lb per lin ft, their weight at each panel point is 13.4 × 17 = 228 lb. An approximate estimate of the weight of the truss and bracing is found in Table 13-2. For a span of 48 ft and a slope of 26½°, the allowance is 7 lb per sq ft of roof surface, or 7 × 114 = 798 lb per panel. The total panel-point load then becomes 7520 + 228 + 798 = 8546 lb, say 8500 lb or 8.5 kips. These panel loads are recorded on the truss diagram of Fig. 13-4a.

13-14 Determination of Stresses

The panel loads of 8500 lb each are shown for the left side of the truss in Fig. 13-4a. Since there are eight panels, the total load on the truss is 8 × 8500 = 68,000 lb. Each end reaction is 68,000 ÷ 2 = 34,000 lb. This truss is symmetrically loaded, so the loads on the right side are not shown. Since the stresses in members similarly located in both halves will be the same, only one half of the stress diagram need be drawn. Note that no load is shown at the heel joint where upper and lower chords meet. Any load here will produce no stress in any truss member, but its magnitude must be considered in the design of the bearing plate or truss connection at the support.

To construct a stress diagram, the first step is to draw the force polygon of the external forces. This polygon for one half the truss consists of the load line *ab*, *bc*, etc., and *ea*, the left reaction. See Fig. 13-4b.

Next, force polygons for the truss members are drawn in the following sequence for the forces about joints *AFE*, *ABGF*, and *FGHE*. Thus, we have established points *f*, *g*, and *h* in the stress diagram. For this type of Fink truss, symmetrically loaded and with equal divisions of the upper chord, each successive member of the upper chord, beginning with the member adjacent to the support, has the same rate of decrease in stress. Referring to the portion of the stress diagram just completed, we note that stresses in *AF* and *BG* have been established; hence the rate of decrease in stress has been established. Members *CJ* and *DK* will have stresses decreased in the same degree. Consequently, we draw a line through points *f* and *g* and call it *X–X*. The points *j* and *k* will be on this line at the intersections, respectively, with lines through c parallel to CJ and through d parallel to DK. The remaining portion of the stress diagram is readily completed. There are other methods of constructing the stress diagram for a Fink truss, but none is constructed more easily than by the method employed here.

Now that the stress diagram is completed, we can determine the magnitudes of the stresses in the members. To do this, we measure the lengths of the lines *in the stress diagram* at the same scale that was used in drawing the force polygon of the external forces. The magnitudes of the stresses are then recorded in Table 13-7. The

TABLE 13-7. Stresses and Sections for Fink Truss

Member	Character of stress	Magnitude of stress (pounds)	Adopted section
AF	Compression	66,600	2 L 4 × 3 × ⁵⁄₁₆
BG	Compression	62,800	2 L 4 × 3 × ⁵⁄₁₆
CJ	Compression	59,000	2 L 4 × 3 × ⁵⁄₁₆
DK	Compression	55,000	2 L 4 × 3 × ⁵⁄₁₆
EF	Tension	59,500	2 L 3½ × 3 × ⁵⁄₁₆
HE	Tension	51,000	2 L 3½ × 3 × ⁵⁄₁₆
LE	Tension	34,000	2 L 2½ × 2 × ⁵⁄₁₆
FG	Compression	7,560	2 L 2½ × 2 × ⁵⁄₁₆
HI	Compression	15,200	2 L 2½ × 2 × ⁵⁄₁₆
JK	Compression	7,560	2 L 2½ × 2 × ⁵⁄₁₆
GH	Tension	8,500	2 L 2½ × 2 × ⁵⁄₁₆
IJ	Tension	8,500	2 L 2½ × 2 × ⁵⁄₁₆
IL	Tension	17,000	2 L 2½ × 2 × ⁵⁄₁₆
KL	Tension	25,000	2 L 2½ × 2 × ⁵⁄₁₆

character of the stress in each member is found by the method described in Art. 13-6 and entered in the second column of Table 13-7.

The eight-panel Fink truss of this example is the same shape as the one shown at the lower right corner of Table 13-6. The reader may be interested in satisfying himself that the same values for stresses in the members will be obtained by the method of stress coefficients discussed in Art. 13-10.

13-15   Identification of Truss Members and Joints

The system of lettering with letters written on each side of the individual members, as shown in Figs. 13-3a and 13-4a, is known as Bow's Notation. It provides a systematic method for identifying each member and force by two letters and is essential when a stress diagram is to be drawn. Another system of notation commonly used for identification of members and joints but not directly related to stress analysis is shown in Fig. 13-6. In this method, letters are placed at the ends of members. For example, the members of the lower chord beginning at the left end of the truss are designated L_0L_1, L_1L_2, and L_2L_3. In this figure, the magnitudes of the stresses have been taken from Table 13-7 and recorded on the truss members, together with C or T to indicate the character of the stress.

13-16   Design of Compression Members

Now that the character and magnitude of the stress in the members have been established, the size of the members can be determined.

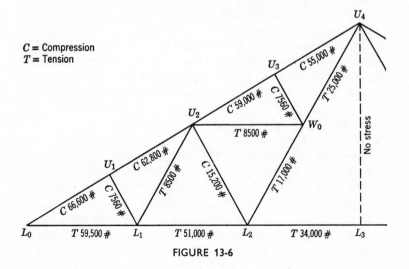

FIGURE 13-6

The customary practice is to use two angles for truss members, separating them by the thickness of the gusset plates required for making connections at the joints (Fig. 13-1). Generally, the most efficient section consists of two angles having unequal legs, with the longer legs placed back-to-back. The compression members are struts, and their load capacities are determined in accordance with the principles explained in Art. 10-10.

First consider the upper chord; it resists compressive stresses. If the length is not too great, it is economical to use one continuous member for the entire length. This avoids the expense of making splices. We see in Fig. 13-6 that the stress in L_0U_1 is 66,600 lb and that the stresses in the other members of the upper chord decrease in magnitude as we approach the apex of the truss. This stress, therefore, will determine the size of the upper chord. The stress is 66,600 lb, its length is 6 ft 8 in., and we can use Table 10-5 to determine its size. Note that this table gives allowable axial loads for double-angle struts, with the long legs back-to-back, separated by $\frac{3}{8}$ in., the thickness of the gusset plate. Here we find that 2 angles $4 \times 3 \times \frac{5}{16}$ will safely support an axial load of 66,000 lb if the length of the member is 8 ft 0 in. Consequently, this section is selected for the entire upper chord. Other sizes might be used, but the cross-sectional area of this member is 4.18 sq in., and it is probably the most economical selection.

Next consider the member U_2L_2. Its length is 6 ft 8 in., and it resists a compressive stress of 15,200 lb. A section composed of 2 angles $2\frac{1}{2} \times 2 \times \frac{5}{16}$ is the minimum size used for any member of the truss regardless of whether the member is in compression or tension. Table 10-5 shows that such a section will support an axial load of 31,000 lb for a length of 7 ft 0 in. and therefore is accepted for member U_2L_2.

The only other members in compression in this half of the truss are members U_1L_1 and U_3W_0. The stress in each of these members is 7560 lb, and the length is 3 ft 4 in. each. Table 10-5 shows that 2 angles $2\frac{1}{2} \times 2 \times \frac{5}{16}$ are acceptable for these two members. We have now determined the size of the compression members for the left half of the truss. The same sizes will be used for similarly located members on the opposite side of the truss. The adopted sections are recorded in Table 13-7.

13-17 Design of Tension Members

As noted in Art. 11-13, all members in direct tension are designed on the basis of the net section, taking the diameter of the hole to be deducted as $\frac{1}{8}$ in. greater than the nominal diameter of the rivets being used. In designing tension members of roof trusses, Table 13-8 will be helpful. Note that this table gives the effective net area of *single* angles with one rivet hole deducted.

The lower chord receives the greatest tensile stresses, and the maximum tensile stress in any member is 59,500 lb, the stress in L_0L_1. Because there will be two angles in each member, each angle in member L_0L_1 must resist a force of 59,500 × $\frac{1}{2}$, or 29,750 lb. The allowable tensile stress of A36 steel is 22,000 psi (Table 3-2). Then 29,750 ÷ 22,000 = 1.35 sq in., the required net area of each angle. The rivets used in the truss will be $\frac{3}{4}$ in. in diameter, and in Table 13-8 we find that the net area of one angle 3 × $2\frac{1}{2}$ × $\frac{5}{16}$ is 1.35 sq in. Consequently, this portion of the lower chord is composed of 2 angles 3 × $2\frac{1}{2}$ × $\frac{5}{16}$. The span of the truss is 48 ft 0 in. If the span is not too great, splices in the lower chord may be avoided by extending the 2 angles 3 × $2\frac{1}{2}$ × $\frac{5}{16}$ over the entire span. In this example, however, we use this section only for the members L_0L_1 and L_1L_2, making a splice at joint L_2.

The stress in L_2L_3 is 34,000 lb. Each angle will resist a force of 34,000 × $\frac{1}{2}$, or 17,000 lb. Then 17,000 ÷ 22,000 = 0.78 sq in., the required net area of each angle. In Table 13-8, we select 2 angles $2\frac{1}{2}$ × 2 × $\frac{5}{16}$, the minimum-size member to be used in the truss.

A continuous member is used for members L_2W_0 and W_0U_4. The greater stress in these two tension members is 25,000 lb. Then, as before, 25,000 × $\frac{1}{2}$ = 12,500 lb and 12,500 ÷ 22,000 = 0.57 sq in., the required net area of the continuous member L_2U_4. The minimum size, 2 angles $2\frac{1}{2}$ × 2 × $\frac{5}{16}$, will be used.

The other tensile members in this half of the truss are L_1U_2 and U_2W_0. The stress to be resisted in each of these members is 8500 lb, and the minimum size, 2 angles $2\frac{1}{2}$ × 2 × $\frac{5}{16}$, is selected for both.

13-18 Rivets in Truss Members

Figure 13-7a shows the truss members in their proper positions. The first step in making such a drawing is to make an accurate

TABLE 13-8. Effective Net Areas of Angles with One Rivet Hole Deducted

Size (inches)	Gross area (sq in.)	Net area (sq in.)		Size (inches)	Gross area (sq in.)	Net area (sq in.)	
		⅞ in. rivets	¾ in. rivets			⅞ in. rivets	¾ in. rivets
2½ × 2 × ³⁄₁₆	0.81		0.64	4 × 3½ × ¼	1.81	1.56	1.59
¼	1.06		0.84	⁵⁄₁₆	2.25	1.94	1.98
⁵⁄₁₆	1.31		1.04	⅜	2.67	2.29	2.34
				⁷⁄₁₆	3.09	2.65	2.70
3 × 2 × ¼	1.19	0.94	0.97	½	3.50	3.00	3.06
⁵⁄₁₆	1.47	1.16	1.20				
⅜	1.73	1.35	1.40	5 × 3 × ¼	1.94	1.69	1.72
				⁵⁄₁₆	2.40	2.09	2.13
3 × 2½ × ¼	1.31	1.06	1.09	⅜	2.86	2.48	2.53
⁵⁄₁₆	1.62	1.31	1.35	⁷⁄₁₆	3.31	2.87	2.92
⅜	1.92	1.54	1.59	½	3.75	3.25	3.31
3½ × 2½ × ¼	1.44	1.19	1.22	5 × 3½ × ¼	2.06	1.81	1.84
⁵⁄₁₆	1.78	1.47	1.51	⁵⁄₁₆	2.56	2.25	2.29
⅜	2.11	1.73	1.78	⅜	3.05	2.67	2.72
⁷⁄₁₆	2.43	1.99	2.04	⁷⁄₁₆	3.53	3.09	3.14
½	2.75	2.25	2.31	½	4.00	3.50	3.56
				⅝	4.92	4.29	4.37
3½ × 3 × ¼	1.56	1.31	1.34				
⁵⁄₁₆	1.93	1.62	1.66	6 × 3½ × ¼	2.31	2.06	2.09
⅜	2.30	1.92	1.97	⁵⁄₁₆	2.87	2.56	2.60
⁷⁄₁₆	2.65	2.21	2.26	⅜	3.42	3.04	3.09
½	3.00	2.50	2.56	½	4.50	4.00	4.06
4 × 3 × ¼	1.69	1.44	1.47	6 × 4 × ⅜	3.61	3.23	3.28
⁵⁄₁₆	2.09	1.78	1.82	⁷⁄₁₆	4.18	3.74	3.80
⅜	2.48	2.10	2.15	½	4.75	4.25	4.31
⁷⁄₁₆	2.87	2.43	2.48	⁹⁄₁₆	5.31	4.74	4.82
½	3.25	2.75	2.81				

dot-and-dash line layout similar to the truss diagram. Now that their sizes have been determined, the members are superimposed on these lines, with the gage lines of the angles coinciding with the lines of the layout.

The rivets that connect the truss members are ¾ in. in diameter. The allowable stresses for the rivets are 15,000 psi for the shearing unit stress and 48,500 psi for the bearing unit stress. These are the stresses used in computing the allowable loads for rivets, as shown in Table 11-2; consequently, this table will be used to determine the number of rivets in the truss members.

For this problem, it is important to realize that a similar condition

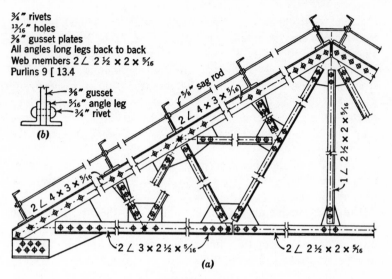

FIGURE 13-7

obtains at the end of each truss member. In each instance, the rivets are $\frac{3}{4}$ in. in diameter, the gusset is $\frac{3}{8}$ in. in thickness, and the legs of the angles are $\frac{5}{16}$ in. thick. See Fig. 13-7b. By consulting Table 11-2, we find the following values.

$$\begin{aligned}
\text{double shear} &= 13.25 \text{ kips} \\
\tfrac{3}{8}\text{-in. bearing} &= 13.7 \text{ kips} \\
\tfrac{5}{16}\text{-in. bearing} \times 2 &= 2 \times 11.4 = 22.8 \text{ kips}
\end{aligned}$$

The smallest and therefore the controlling value is 13.25 kips or 13,250 lb.

It is customary to use not less than two rivets at the end of any truss member, regardless of the magnitude of the stress to be transferred. The minimum pitch of the rivets is three times the rivet diameter, or $2\frac{1}{2}$ in. The maximum pitch is 6 in. Frequently the size of the gusset plate requires a greater number of rivets than are needed to transfer the loads in the members.

Consider first the member L_0U_1 at its connection to the gusset plate at the support. The stress in this member is 66,600 lb and, since the controlling value of one rivet is 13,250 lb, 66,600 ÷ 13,250 = 5+. We use six rivets. The stress in L_0L_1 is 59,500; hence, 59,500 ÷ 13,250 = 4+. Therefore, we use five rivets. See Fig. 13-7.

The members L_0L_1 and L_1L_2 are one continuous section. The stresses in these members are 59,500 lb and 51,000 lb, respectively. Then 59,500 − 51,000 = 8500 lb, the stress to be transferred by rivets in the gusset, Thus, 8500 divided by 13,250 is less than one rivet, but three rivets are used at joint L_1 because of the size of the gusset plate.

The stress in member L_1L_2 is 51,000 lb. This member ends at joint L_2. Consequently, 51,000 ÷ 13,250 = 3+ and four rivets are used at joint L_2.

The stress in member U_1L_1 at joint L_1 is 7560 lb. Then 7560 divided by 13,250 is less than one, and we use two, the minimum number of rivets. For the same reason, two rivets are used in this member at joint U_1.

The stresses in U_1U_2 and U_2U_3 are 62,800 lb and 59,000 lb, respectively. Hence, 62,800 − 59,000 = 3800 lb, the stress to be transferred from one side of the joint to the other. Then 3800 divided by 13,250 is less than one rivet. At this joint, several members come together; and because of the size of the gusset plate, five rivets are used, as shown in Fig. 13-7.

The number of rivets in the other truss members are determined in a similar manner; they are shown in Fig. 13-7a. Attention is called to members U_4L_3, the vertical web member at the mid-span of the truss. No stress in this member results from the panel loads. However, a single angle is used to resist the tendency of member L_2L_3 to sag, or deflect under its own weight.

Not shown in Fig. 13-7 are the intermittent rivets and fillers required in double-angle compression members at the spacings specified in Art. 10-10. Similar intermittent connections are used in double-angle tension members to reduce their flexibility, with a maximum spacing of 3 to $3\frac{1}{2}$ ft.

13-19 Welded Truss Connections

Welds may be substituted for rivets in gusset-type connections such as those used for the truss shown in Fig. 13-7. When this is done, the design procedure is similar to that employed in Example 2 of Art. 12-5 except that two angles are involved, each carrying half the total stress in the member. Figure 12-6b would then show angles welded to both sides of the plate. However, in order to realize the full effectiveness of welding, the chord members may be made of structural tees (Art. 1-6), and the double-angle web members may be welded directly to each side of the stem of the tee section. Figure 12-9k shows such an arrangement.

Figure 13-8a shows a portion of a parallel-chord truss. Consider the joint L_4. Here we have the web members U_5L_4, a vertical

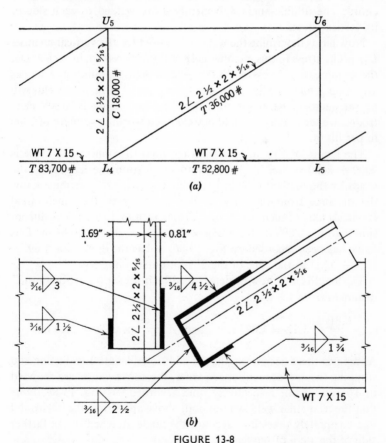

(a)

(b)

FIGURE 13-8

compression member in which the stress is 18,000 lb, and U_6L_4, a sloping member which resists a tensile force of 36,000 lb. Each of these two members consists of 2 angles $2\frac{1}{2} \times 2 \times \frac{5}{16}$. If the span of the truss is not too great, it is economical to use one continuous section for the entire lower chord. The greatest stress in the lower chord is in L_3L_4, a tensile stress whose magnitude is 83,700 lb. If F_t, the allowable tensile stress, is 22,000 psi (Table 3-2), 83,700 ÷ 22,000 = 3.8 sq in., the required net area of the member L_3L_4. The area of a WT 7 × 15 is 4.42 sq in. This tee section is cut from a W 14 × 30 and is used as a continuous member for the entire lower chord. The double-angle web members are welded to each side of its vertical stem.

Now let us determine the welds to be used for the vertical member U_5L_4. The stress in this member is 18,000 lb; therefore, in each angle, the compressive stress is 18,000 ÷ 2 = 9000 lb. Since the angles are $\frac{5}{16}$ in. thick, we find from Table 12-2 that $\frac{3}{16}$-in. fillet welds will be satisfactory. Referring to Table 12-1 and using E 70 XX electrodes, we see that a $\frac{3}{16}$-in. fillet weld has a working strength of 2800 lb per lin in.

The length of $\frac{3}{16}$-in. fillet weld required to connect one angle is 9000 ÷ 2800 = 3.22 in. In order to proportion the lengths of the welds by the method used in Example 2 of Art. 12-5, we must know the distance from the back of the short leg to the longitudinal (vertical) axis of member U_5L_4. This is given in Table 1-5 as dimension y and is 0.809 in. for a $2\frac{1}{2} \times 2 \times \frac{5}{16}$ angle. Use 0.81 in. This value has been recorded on Fig. 13-8b, as has the dimension 1.69 in. (2.5 − 0.81) to the edge of the $2\frac{1}{2}$-in. outstanding leg.

The lengths may now be determined by inverse proportion of their distances from the axis:

$$\frac{0.81}{2.5} \times 3.22 = 1.05 \text{ in.} \quad \text{and} \quad \frac{1.69}{2.5} \times 3.22 = 2.18 \text{ in.}$$

Adding to these design lengths the additional length for starting and stopping the arc, as mentioned earlier, the above values are rounded to $1\frac{1}{2}$ and 3 in., respectively, and recorded on Fig. 13-8b. Note that the triangular weld symbol both above and below the horizontal line adequately takes into account the angle attached to the farther side of the stem of the tee (see Art. 12-9).

The stress in member L_4U_6 is 36,000 lb.; therefore, the stress in each of the two angles is $36,000 \div 2 = 18,000$ lb. The length of fillet weld required to connect one angle is $18,000 \div 2800 = 6.44$ in. To determine the lengths of the individual welds,

$$\frac{0.81}{2.5} \times 6.44 = 2.09 \text{ in.} \quad \text{and} \quad \frac{1.69}{2.5} \times 6.44 = 4.35 \text{ in.}$$

Adding approximately ¼ in. for starting and stopping the arc, these lengths become 2½ and 4¾ in., respectively. There is ample room on the tee stem to accommodate welds of these lengths. However, the connection of L_4U_6 to the stem in Fig. 13-8 has been used to illustrate the situation where space is restricted, and the weld lengths shown do not apply to the foregoing computations. If there is insufficient room for the two lengths required, they may be reduced by 2½ in. by running a fillet weld across the end of the angle. This is the configuration of the weld shown in the figure.

13-20   End Bearing and Anchorage

Heel joint is the name given to the joint of a truss where the upper and lower chords meet at the support. Typical joints for trusses resting on masonry walls are shown in Fig. 13-9. A typical form of connection, as shown in Fig. 13-9*b* and *c*, consists of *shoe angles* riveted to the gusset plate to permit the center line of the support to meet the intersection of the working lines of the upper and lower chords. The rivets in the angles over the bearing should be of sufficient number to transfer the end reaction of the truss to the bearing plate. In addition, the net section of the gusset plate should

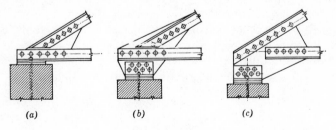

(*a*)   (*b*)   (*c*)

FIGURE 13-9

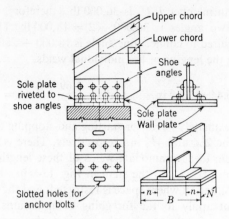

FIGURE 13-10

be large enough to resist the vertical shear, which is equal in magnitude to the reaction.

Fundamentally, the size of the bearing plate on a masonry wall depends upon the magnitude of the reaction and the allowable bearing pressure on the masonry. It is considered good practice, however, to extend the end of the truss at least 6 in. beyond the center line of the bearing to provide a minimum bearing length of 12 in. To help distribute the truss more uniformly, a *sole plate* is riveted to the lower chord or shoe angles, the sole plate in turn resting upon the bearing plate (Fig. 13-10). Both the sole plate and the bearing plate should have minimum thicknesses of ½ in. They should not project more than 3 or 4 in. beyond the sides of the angles, and slotted holes should be provided for anchor bolts. Owing to expansion and contraction, the slotted holes should be sufficiently long to provide for a movement of ⅛ in. for every 10 ft of truss span. Anchor bolts of ⅝-in. diameter are commonly used. The slotted holes through which they pass should be at least ⁵⁄₁₆ in. greater in width that the diameter of the bolts. For spans exceeding about 80 ft, a rocker or roller bearing should be employed.

14

Theory of
Plastic
Design

‖‖

14-1 Elastic Design

The discussions up to this point relating to the design of members in
bending have been based on bending stresses well within the yield
point stress. The allowable stresses are based on the *theory of elastic
design*. However, it has been found by tests that members can carry
loads much higher than anticipated even when the yield point stress
is reacted at sections of maximum bending moment. This is particularly evident in continuous beams and structures employing rigid
frames. An inherent property of structural steel is its ability to resist
large deformations without failure. The deformations in these
structures fall within the *plastic range*, occurring without an increase
in bending stress. Because of this phenomenon, the *plastic design
theory*, sometimes called the *ultimate strength design theory*, has
been developed.

The 1969 AISC Specification permits the use of plastic design
with the following ASTM steels: A36, A242, A441, A529, A572,
A588. See Art. 3-1 for the names of these steels.

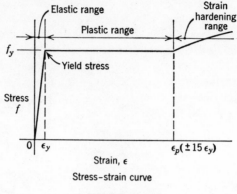

Stress–strain curve

FIGURE 14-1

14-2 Stress-Strain Diagram

Figure 14-1 is a graphic representation of a steel specimen as it deforms under stress. The horizontal scale is greatly exaggerated. The term *strain* is synonymous with *deformation*. The purpose of this graph is to show that, up to a stress f_y, the yield point, the deformations are directly proportional to the applied stresses and that, beyond this point of maximum elastic strain, there is a deformation without an increase in stress. The total deformation is approximately 15 times that produced elastically. This deformation is called the *plastic range*, beyond which *strain hardening* begins when further deformation can occur only with an increase in stress.

14-3 Plastic Moment, Plastic Hinge

Article 6-3 explains the design of members in bending in accordance with the theory of elasticity. When the extreme fiber stress does not exceed the elastic limit, the bending stresses in the cross section of a beam are directly proportional to their distances from the neutral surface. In addition, the strains (deformations) in these fibers are also directly proportional to their distances from the neutral surface. Both stress and strain are zero at the neutral surface, and both increase to maximum magnitudes at the fibers farthest from the neutral surface.

The following example, which illustrates the design of a steel beam for bending, employs the theory of *elastic design*.

A simple steel beam has a span of 16 ft 0 in. with a concentrated load of 18,000 lb at the center of the span. The section used for this load is an S 12 × 31.8, the beam is adequately braced throughout its length, and the weight of the beam is ignored in the computations. Let us compute the maximum extreme fiber stress.

To do this, we use the flexure formula

$$f = \frac{M}{S} \qquad \text{(Art. 5-2)}$$

Then

$$M = \frac{Pl}{4} = \frac{18{,}000 \times 16 \times 12}{4} = 864{,}000 \text{ in-lb}$$

which is the maximum bending moment. Referring to Table 8-2, we see that $S = 36.4$ in.^{3} for an S 12 × 31.8. Thus,

$$f = \frac{M}{S} \quad \text{and} \quad f = \frac{864{,}000}{36.4} = 23{,}800 \text{ say } 24{,}000 \text{ psi}$$

which is the actual fiber stress on the fiber farthest from the neutral surface. See Fig. 14-2*d*. Note that this stress occurs only at a beam section at the center of the span where the bending moment has its maximum value. Figure 14-2*e* shows ϵ, the deformation on the fiber farthest from the neutral surface. Note that both the stresses and deformations are directly proportional to their distances from the neutral surface in *elastic design*.

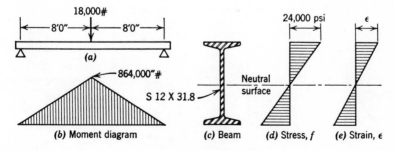

(a)

(b) Moment diagram   (c) Beam   (d) Stress, f   (e) Strain, ϵ

FIGURE 14-2

When a beam is loaded to produce an extreme fiber stress in excess of the yield point, the property of ductility affects the distribution of the stresses. *Ductility* of a material is the property that permits it to undergo plastic deformation when subjected to stresses. Structural steel is a ductile material. Ductility permits a redistribution of stresses in a beam cross section. Fibers that were less stressed originally come to the assistance of the more highly stressed fibers.

Assume that the bending moment on a beam is of such magnitude that the extreme fiber stress is f_y, the yield stress. Then, if M_y is the elastic bending moment at the yield stress, $M = M_y$, and the distribution of the stresses in the cross section is as shown in Fig. 14-3a; the maximum bending stress f_y is at the extreme fiber.

Next consider that the loading and the resulting bending moment have been increased. M is now greater than M_y. The stress on the extreme fiber is still f_y, but *the material has yielded* and a greater area of the cross section is also stressed to f_y. The stress distribution is shown in Fig. 14-3b.

Again imagine the load to be further increased. The stress on the extreme fiber is still f_y and, theoretically, *all fibers in the cross section are stressed to f_y*. This is an idealized plastic stress distribution; it is shown in Fig. 14-3d. The bending moment producing this condition is M_p, the plastic bending moment. In reality, 10% of the central portion of the cross section resists elastically distributed stresses, as indicated in Fig. 14-3c. This small area is considered to be negligible, and we assume that the stresses on all the fibers of the cross section are f_y, as shown in Fig. 14-3d. The section is now said

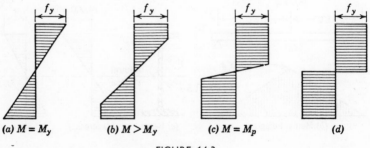

(a) $M = M_y$ (b) $M > M_y$ (c) $M = M_p$ (d)

FIGURE 14-3

The following example, which illustrates the design of a steel beam for bending, employs the theory of *elastic design*.

A simple steel beam has a span of 16 ft 0 in. with a concentrated load of 18,000 lb at the center of the span. The section used for this load is an S 12 × 31.8, the beam is adequately braced throughout its length, and the weight of the beam is ignored in the computations. Let us compute the maximum extreme fiber stress.

To do this, we use the flexure formula

$$f = \frac{M}{S} \qquad \text{(Art. 5-2)}$$

Then

$$M = \frac{Pl}{4} = \frac{18,000 \times 16 \times 12}{4} = 864,000 \text{ in-lb}$$

which is the maximum bending moment. Referring to Table 8-2, we see that $S = 36.4$ in.³ for an S 12 × 31.8. Thus,

$$f = \frac{M}{S} \qquad \text{and} \qquad f = \frac{864,000}{36.4} = 23,800 \text{ say } 24,000 \text{ psi}$$

which is the actual fiber stress on the fiber farthest from the neutral surface. See Fig. 14-2d. Note that this stress occurs only at a beam section at the center of the span where the bending moment has its maximum value. Figure 14-2e shows ϵ, the deformation on the fiber farthest from the neutral surface. Note that both the stresses and deformations are directly proportional to their distances from the neutral surface in *elastic design*.

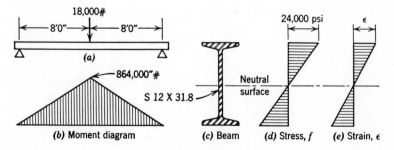

(a)

(b) Moment diagram 864,000"#

S 12 X 31.8 Neutral surface

(c) Beam (d) Stress, f (e) Strain, ε

18,000# 8'0" 8'0" 24,000 psi ε

FIGURE 14-2

When a beam is loaded to produce an extreme fiber stress in excess of the yield point, the property of ductility affects the distribution of the stresses. *Ductility* of a material is the property that permits it to undergo plastic deformation when subjected to stresses. Structural steel is a ductile material. Ductility permits a redistribution of stresses in a beam cross section. Fibers that were less stressed originally come to the assistance of the more highly stressed fibers.

Assume that the bending moment on a beam is of such magnitude that the extreme fiber stress is f_y, the yield stress. Then, if M_y is the elastic bending moment at the yield stress, $M = M_y$, and the distribution of the stresses in the cross section is as shown in Fig. 14-3a; the maximum bending stress f_y is at the extreme fiber.

Next consider that the loading and the resulting bending moment have been increased. M is now greater than M_y. The stress on the extreme fiber is still f_y, but *the material has yielded* and a greater area of the cross section is also stressed to f_y. The stress distribution is shown in Fig. 14-3b.

Again imagine the load to be further increased. The stress on the extreme fiber is still f_y and, theoretically, *all fibers in the cross section are stressed to f_y.* This is an idealized plastic stress distribution; it is shown in Fig. 14-3d. The bending moment producing this condition is M_p, the plastic bending moment. In reality, 10% of the central portion of the cross section resists elastically distributed stresses, as indicated in Fig. 14-3c. This small area is considered to be negligible, and we assume that the stresses on all the fibers of the cross section are f_y, as shown in Fig. 14-3d. The section is now said

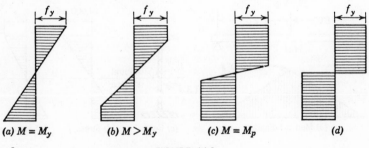

(a) $M = M_y$ (b) $M > M_y$ (c) $M = M_p$ (d)

FIGURE 14-3

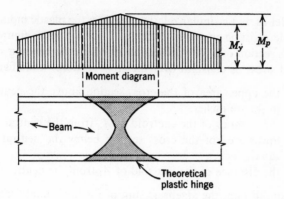

FIGURE 14-4

to be fully plastic, and a further increase in load will result in large deformations, the beam acting as if it were hinged at this section. We call this a *plastic hinge* at which free rotation is permitted only after M_p has been attained. See Fig. 14-4. At sections of the beam in which this condition prevails, the bending resistance of the cross section has been exhausted.

14-4   Plastic Section Modulus

In elastic design, the moment producing the maximum allowable resisting moment may be found by the flexure formula

$$M = f \times S$$

in which $M =$ the maximum allowable bending moment in inch-pounds

$f =$ the maximum allowable bending stress in pounds per square inch

$S =$ the section modulus in inches to the third power

If the extreme fiber is stressed to the yield stress,

$$M_y = f_y \times S$$

in which $M_y =$ the elastic bending moment at yield stress

$f_y =$ the yield stress in pounds per square inch

$S =$ the section modulus in inches to the third power

Now let us find a similar relation between the plastic moment and its plastic resisting moment. Refer to Fig. 14-5. This figure shows the cross section of a W or S section in which the bending stress f_y, the yield stress, is constant over the cross section. In the figure,

A_u = the upper area of the cross section above the neutral axis in square inches

y_u = the distance of the centroid of A_u from the neutral axis

A_l = lower area of the cross section below the neutral axis in square inches

y_l = the distance of the centroid of A_l from the neutral axis

For equilibrium, the algebraic sum of the horizontal forces must be zero. Then

$$\sum H = 0$$

or

$$[A_u \times (+f_y)] + [A_l \times (-f_y)] = 0$$

and

$$A_u = A_l$$

This shows that the neutral axis divides the cross section into equal areas, which is apparent in symmetrical sections, but it applies to unsymmetrical sections as well. Also, the bending moment equals the sum of the moments of the stresses in the section. Thus, for M_p, the plastic moment,

$$M_p = (A_u \times f_y \times y_u) + (A_l \times f_y \times y_l)$$

or

$$M_p = f_y[(A_u \times y_u) + (A_l \times y_l)]$$

and

$$M_p = f_y \times Z$$

The quantity $(A_u y_u + A_l y_l)$ is called the *plastic section modulus* of the cross section, and it is designated by the letter Z. Since it is an area multiplied by a distance, it is in units to the third power. If the area is in units of square inches and the distance is in linear inches, Z, the section modulus, is in units of inches to the third power.

The plastic section modulus is always larger than the elastic section modulus.

It is important to note that, in plastic design, the neutral axis for unsymmetrical cross sections does not pass through the centroid of the section. In plastic design, the neutral axis divides the cross section into *equal* areas.

14-5 Computation of Plastic Section Moduli

The notation used in Art. 14-4 is appropriate for both symmetrical and unsymmetrical sections. Consider now a symmetrical section such as a W or S shape, as shown in Fig. 14-5. $A_u = A_l$, $y_u = y_l$ and $A_u + A_l = A$, the total area of the cross section Then

$$M_p = (A_u \times f_y \times y_u) + (A_l \times f_y \times y_l)$$

and

$$M_p = f_y \times A \times y \quad \text{or} \quad M_p = f_y \times Z$$

in which f_y = the yield stress in pounds per square inch

A = the total area of the cross section in square inches

y = the distance from the neutral axis to the centroid of the portion of the area on either side of the neutral axis in inches

Z = the plastic section modulus in inches to the third power.

Now since $Z = A \times y$, we can readily compute the value of the plastic section modulus of a given cross section.

Consider a W 16 × 45. Referring to Table 1-1, we find that its total depth is 16.12 in. and its cross-sectional area is 13.3 sq in. From the AISC Manual tables giving properties for designing structural tees, we find that a WT 8 × 22.5 (which is cut from a

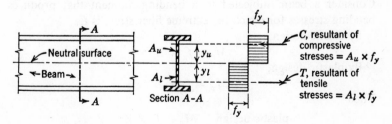

FIGURE 14-5

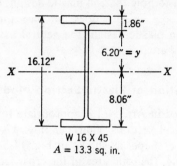

W 16 X 45
$A = 13.3$ sq. in.

FIGURE 14-6

W 16 × 45) has its centroid located 1.86 in. from the top face of the tee. Therefore, the distance from the centroid of the upper half of the W 16 × 45 to its extreme fiber is also 1.86 in. This dimension is recorded in Fig. 14-6. Subtracting 1.86 from half the beam depth, 8.06 in., the distance y in Fig. 14-6 becomes 6.20 in. Then the plastic section modulus of the W 16 × 45 is

$$Z = A \times y = 13.3 \times 6.20 = 82.1 \text{ in.}^3$$

The plastic section moduli of symmetrical sections are computed in this manner, and Table 14-1 gives values of Z for certain selected W and S sections. The ratio d/t_w given in the table is the depth-thickness ratio of the beam webs and is used in the formulas referred to in the footnote to the table.

14-6 Shape Factor

Consider a beam subjected to a bending moment that produces bending stresses for which the extreme fiber stress is f_y.

By elastic design,

$$M_y = f_y \times S$$

By plastic design,

$$M_p = f_y \times Z$$

Then

$$\frac{\text{plastic design}}{\text{elastic design}} = \frac{M_p}{M_y} = \frac{f_y \times Z}{f_y \times S} = \frac{Z}{S}$$

TABLE 14-1. Typical Plastic Design Selection Table**
for Shapes Used as Beams or Columns

$$Z_X$$

Z_x	Shape	A	$\dfrac{d}{t_w}$	r_x	r_y	$F_y = 36$ ksi		$F_y = 50$ ksi	
						M_p	P_y	M_p	P_y
In.³		In.²		In.	In.	Kip-ft.	Kips	Kip-ft.	Kips
152	W 24 × 61	18.0	56.6	9.25	1.38	456	* 648	633	* 900
144	W 21 × 62	18.3	52.5	8.54	1.77	432	* 659	600	* 915
140	W 12 × 92	27.1	23.2	5.40	3.08	420	976	—	—
138	S 20 × 65.4	19.2	40.0	7.84	1.19	414	691	575	* 960
126	W 21 × 55	16.2	55.5	8.40	1.73	378	* 583	—	—
126	W 14 × 74	21.8	31.5	6.05	2.48	378	785	525	1090
125	S 18 × 70	20.6	25.3	6.71	1.08	375	742	521	1030
123	W 18 × 60	17.7	43.9	7.47	1.68	369	* 637	513	* 885
119	W 12 × 79	23.2	26.3	5.34	3.05	357	835	—	—
118	W 16 × 64	18.8	36.1	6.66	1.97	354	677	492	940
115	W 14 × 68	20.0	33.6	6.02	2.46	345	720	479	1000
114	W 10 × 89	26.2	17.7	4.55	2.63	342	943	475	1310
112	W 18 × 55	16.2	46.5	7.42	1.67	336	* 583	467	* 810
108	W 21 × 49	14.4	56.6	8.21	1.31	324	* 518	450	* 720
106	W 16 × 58	17.1	39.0	6.62	1.96	318	616	442	* 855
105	S 18 × 54.7	16.1	39.0	7.07	1.14	315	580	438	* 805
102	W 14 × 61	17.9	36.8	5.98	2.45	306	644	—	—
101	W 18 × 50	14.7	50.3	7.38	1.65	303	* 529	421	* 735
97.8	W 10 × 77	22.7	19.9	4.49	2.60	293	817	408	1140
95.3	W 21 × 44	13.0	59.4	8.07	1.27	286	* 468	—	—
91.8	W 16 × 50	14.7	42.8	6.68	1.59	275	529	383	* 735
90.6	W 10 × 72	21.2	20.6	4.46	2.59	272	763	378	1060
89.7	W 18 × 45	13.2	53.3	7.30	1.62	269	* 475	— .	
87.1	W 14 × 53	15.6	37.7	5.90	1.92	261	562	363	* 780
86.5	W 12 × 58	17.1	34.0	5.28	2.51	260	616	—	—
82.8	W 10 × 66	19.4	22.7	4.44	2.58	248	698	345	970
82.1	W 16 × 45	13.3	46.6	6.64	1.57	246	* 479	342	* 665
78.4	W 18 × 40	11.8	56.6	7.21	1.27	235	* 425	327	* 590
78.4	W 14 × 48	14.1	40.7	5.86	1.91	235	508	327	* 705
77.1	S 15 × 50	14.7	27.3	5.75	1.03	231	529	321	735
75.0	W 10 × 60	17.7	24.7	4.41	2.57	225	637	—	—
72.8	W 16 × 40	11.8	52.1	6.62	1.56	218	* 425	303	* 590
72.5	W 12 × 50	14.7	32.9	5.18	1.96	218	529	302	735

* Check shape for compliance with Formulas (2.7-1a) or (2.7-1b), Section 2.7, AISC
Specification, as applicable, when subjected to combined axial force and bending
moment at ultimate loading.
** Compiled from data in the more complete tables contained in the 7th Edition of
the *Manual of Steel Construction*. Courtesy American Institute of Steel Construction.

The relation between the two moments, $\dfrac{M_p}{M_y}$ is called the *shape factor*. It is represented by the letter u. Thus,

$$u = \frac{Z}{S}$$

Let us compute the value of u, the shape factor for a rectangle whose width is b and whose depth is d. See Fig. 14-7.

The elastic section modulus of this rectangle about the neutral axis parallel to the base is

$$S = \frac{bd^2}{6} \qquad \text{(Table 5-1)}$$

For the plastic section modulus of the rectangle,

$$Z = \left(b \times \frac{d}{2} \times \frac{d}{4}\right) + \left(b \times \frac{d}{2} \times \frac{d}{4}\right)$$

or

$$Z = \frac{bd^2}{8} + \frac{bd^2}{8} = \frac{bd^2}{4}$$

Then

$$\frac{Z}{S} = \frac{bd^2}{4} \div \frac{bd^2}{6} = \frac{bd^2}{4} \times \frac{6}{bd^2} = \frac{3}{2} = 1.5$$

which is the shape factor for rectangles. Thus, M_p, the plastic moment for a rectangular cross section, is 50% greater than M_y, the elastic moment at yield stress.

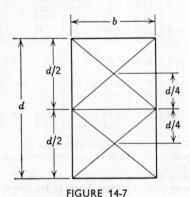

FIGURE 14-7

For the commonly used steel sections, such as W and S sections, the shape factor is approximately 1.12 for bending about the strong axis of the section. This means that these sections can support approximately 12% more than we expect them to carry on the basis of elastic design.

14-7   Restrained Beams

Figure 14-8a shows a uniformly distributed load of w lb per lin ft on a beam that is fixed (restrained from rotation) at both ends. The maximum bending moment for this condition is $wl^2/12$; it occurs at the supports and is a negative quantity. At the center of the span, the moment is positive, and its magnitude is $wl^2/24$.

Consider now that the load is increased from w to w_y, thus producing a stress of f_y in the extreme fibers of the cross section over the supports. At the supports,

$$M_y = -\frac{w_y l^2}{12}$$

and at the center of the span,

$$M_y = +\frac{w_y l^2}{24}$$

as shown in Fig. 14-8b. This is the limit of the elastic behavior; the moments at the supports are still twice the magnitude of the moment at the mid-span at which none of the fibers of the cross section has reached the yield stress.

Now let us further increase the load to w_p lb per lin ft. See Fig. 14-8c. The stresses in the fibers of the cross section *over the supports* will now begin to yield until *all* the fibers are stressed to f_y and plastic hinges are formed at the ends. The beam can no longer resist further rotation at its ends, and the increase in load must be resisted by sections of the beam that are less stressed. The critical section lies at the center of the span. If the load w_p has been increased so that a plastic hinge is formed at the mid-span, the beam will become a *mechanism*. A mechanism exists when sufficient plastic hinges have been formed so that no further loading may be supported by the member. Note that the beam when thus loaded produces three

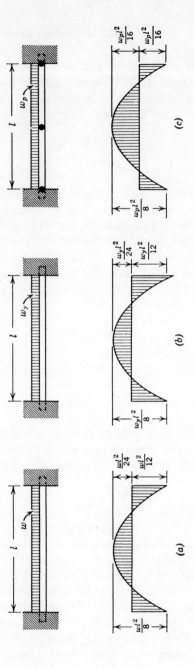

FIGURE 14-8

plastic hinges, one at the mid-span and two at the supports. They are indicated in Fig. 14-8c by the three small solid circles. The moments at the supports and at mid-span are now equal and have magnitudes of $w_p l^2/16$, which means that the full strength of the beam is utilized at three sections instead of two. A load whose magnitude is greater than w_p lb per lin ft would result in a permanent deformation and consequently a practical failure.

14-8 Load Factor

Consider a simple beam of A36 steel laterally supported throughout its length. Its span is 24 ft, and it carries a concentrated load of 42,000 lb at the center of the span. Determine the size of the beam in accordance with the theory of elastic behavior.

The maximum bending moment for this beam is $Pl/4$. Then, since the flexure formula is $S = M/f$,

$$S = \frac{42,000 \times 24 \times 12}{4 \times 24,000} = 126 \text{ in.}^3$$

which is the required section modulus. Table 8-2 shows that a W 21 × 62 has a section modulus of 127 in.³ and is acceptable.

Now let us compute for this beam the magnitude of the concentrated load at the center of the span that would produce a bending moment equal to the plastic resisting moment.

The shape factor for W and S shapes is 1.12; and since $M_p = f_y \times S$, $M_p = 1.12 \times 36,000 \times 126.4 = 5,100,000$ in-lb, the plastic bending moment. Then, to find the load corresponding to this moment since $M = Pl/4$,

$$5,100,000 = \frac{P \times 24 \times 12}{4} \quad \text{and} \quad P = 71,000 \text{ lb}$$

the magnitude of the concentrated load. This load would produce a plastic hinge at the center of the span, and a slight increase in load would result in failure.

The term *load factor* is given to the ratio of the ultimate load to the design load. In this example, it is 71,000/42,000, or 1.7, the load factor.

Referring to Fig. 14-8c, we read in Art. 14-7 that a load greater than w_p lb per lin ft would result in a permanent deformation and a practical failure. Because we cannot determine loads with minute accuracy and because our materials are not always completely uniform, it would be dangerous to allow the actual load to become as large as w_p lb per lin ft. In elastic design, the allowable bending stress, F_b, is decreased to a fraction of F_y, the yield stress. For compact sections, $F_b = 0.66\,F_y$. Therefore, the factor of safety against yielding is $1/0.66$, or 1.5.

In plastic design, it is simpler to base computations on F_y, the yield stress, and to increase the load causing it to be a multiple of the actual load. Here we employ the load factor. For simple and continuous beams, the load factor is 1.7; for rigid frames, it is 1.85.

14-9 Design of Simple Beams

The design of simple beams by either the elastic theory or the plastic theory will usually result in the same size beam, as illustrated in the following two examples.

Example 1. A simple beam of A36 steel has a span of 20 ft and supports a uniformly distributed load of 4800 lb per lin ft, including its own weight. Design this beam in accordance with the *elastic theory*, assuming that the beam is laterally supported throughout its length.

Solution: The maximum bending moment is $wl^2/8$. Then

$$M = \frac{4800 \times 20 \times 20 \times 12}{8} = 2{,}880{,}000 \text{ in-lb}$$

The allowable bending stress, F_b, for a compact beam is 24,000 psi. Then

$$S = \frac{M}{F_b} = \frac{2{,}880{,}000}{24{,}000} = 120 \text{ in.}^3$$

which is the required section modulus. In Table 8-2, select a W 21 × 62; its section modulus is 127 in.^{3}

Example 2. Design the same beam in accordance with the *plastic theory*.

Solution: $w_p = w \times$ load factor. In accordance with Art. 14-8, the load factor for simple beams is 1.7. Then

$$w_p = 4800 \times 1.7 \qquad \text{and} \qquad w_p = 8{,}160 \text{ lb per lin ft}$$

$$M_p = \frac{w_p l^2}{8} = \frac{8160 \times 20 \times 20 \times 12}{8} = 4{,}896{,}000 \text{ in-lb}$$

which is the plastic bending moment.

$$M_p = f_y \times Z \qquad \text{or} \qquad Z = \frac{M_p}{f_y} \qquad \text{(Art. 14-4)}$$

Then

$$Z = \frac{4{,}896{,}000}{36{,}000} = 136 \text{ in.}^3$$

which is the minimum plastic section modulus. In Table 14-1, select a W 21 × 62 in which $Z = 144$ in.^{3}

Notice that elastic theory and the plastic theory have given the same section for this simple beam.

14-10 Design of a Beam with Fixed Ends

A beam that has both ends fixed or restrained is similar to an interior span of a fully continuous beam; the magnitudes of the bending moments are the same. In structural steel, it is economical to design such beams in accordance with the plastic theory because there is a saving of material.

Example 1. A beam of A36 steel with both ends fixed has a clear span of 20 ft. There is a uniformly distributed load of 4800 lb per lin ft, including its own weight. (This is the same loading as that used in the examples of Art. 14-9 for a simply supported beam of the same span.) Design the beam in accordance with the *elastic theory*, assuming full lateral support is provided.

Solution: Referring to Fig. 14-8a, we see that the maximum bending moment is at the supports, a negative moment of $wl^2/12$. Then

$$M = \frac{4800 \times 20 \times 20 \times 12}{12} = 1{,}920{,}000 \text{ in-lb}$$

For a compact section, $F_b = 24,000$ psi. Then

$$S = \frac{M}{F_b} \quad \text{and} \quad S = \frac{1,920,000}{24,000} = 80 \text{ in.}^3$$

which is the required section modulus. Referring to Table 8-2, we find that a W 18×50 has a section modulus of 89.1 in.³ and is acceptable.

Example 2. Design the same beam with fixed ends in accordance with the *plastic theory*.

Solution: $w_p = w \times$ load factor. In Art. 14-8, we found that the load factor for continuous beams is 1.7. Then

$$w_p = 4800 \times 1.7 = 8160 \text{ lb per lin ft}$$

Both the positive and negative bending moments for a continuous beam with loads producing plastic hinges at the supports as well as at the mid-span is

$$M_p = \frac{w_p l^2}{16} = \frac{8160 \times 20 \times 20 \times 12}{16} = 2,448,000 \text{ in-lb} \quad \text{(Fig. 14-8}c\text{)}$$

Then

$$Z = \frac{M_p}{f_y} = \frac{2,448,000}{36,000} = 68 \text{ in.}^3$$

which is the required plastic section modulus. On referring to Table 14-1, we see that a W 16×40 has a plastic section modulus of 72.8 in.³ and is acceptable.

Note that a W 18×50 was required when this restrained beam was designed in accordance with the elastic theory in Example 1 of this article. Therefore, a saving in steel of 10 lb per lin ft was achieved in this case by use of the plastic theory.

Problem 14-10-A. A beam of A36 steel has a span of 22 ft and is fixed at both ends. The load on the beam is 5000 lb per lin ft, including its own weight, and the beam is supported laterally throughout its length. Determine, in accordance with the elastic theory, the lightest weight beam listed in Table 8-2 that will support this load.

Problem 14-10-B. Using the data given for the beam of Problem 14-10-A, determine the lightest weight beam in accordance with the plastic theory.

14-11   Scope of Plastic Design

The purpose of this brief chapter is to explain how the theory of plastic design in steel has evolved. The illustrative problems relate only to beams that are fixed at both ends; they show how the application of plastic design theory may result in economy of material. The theory, however, has applications far beyond fixed beams, particularly in the design of rigid frames where combinations of bending and direct stress are involved. Since the design is based on higher stresses, adequate attention must be given to the possibility of local buckling, and care must be taken to prevent excessive deflections that might occur because of the smaller sections usually resulting from use of the theory. The reader who wishes to pursue the subject of plastic design further is referred to *Commentary on Plastic Design in Steel* by the American Society of Civil Engineers, or *Plastic Design in Steel* by the American Institute of Steel Construction.

Answers to Selected Problems

The answers given below are for those problems marked with an asterisk (*) in the text. They are the result of slide-rule computations carried to three significant figures, except in certain cases where extension of the numerical answer seemed desirable as an aid in interpreting the result.

Chapter 2

2-3-D $f_t = 23,900$ psi

2-7-B $E = 29,400,000$ psi

2-9-A $f_t = 28,200$ psi; No

2-9-E $e = 0.151$ in.

Chapter 4

4-6-C $R_1 = 2620$ lb, $R_2 = 2980$ lb

4-6-E $R_1 = 4100$ lb, $R_2 = 8000$ lb

4-9-D $R_1 = 14,000$ lb, $R_2 = 10,000$ lb. Shear: at left end = 14,000 lb; 4 ft from left end = 10,000 lb and 2000 lb; at right end = −10,000 lb. V = zero at 6 ft from left end.

4-9-E $R_1 = 5340$ lb, $R_2 = 2660$ lb. Shear: at left end = zero; 4 ft from left end = -2000 lb and 3340 lb; at right end = -2660 lb. V = zero at 10.68 ft from left end.

4-12-A   Bending moment: at left end = zero; 3 ft from left end = 31,200 ft-lb; 10 ft from left end = 48,000 ft-lb; at right end = zero.

4-19-C $R_1 = 6700$ lb, $R_2 = 10,300$ lb. Shear: at left end = 6700 lb; 5 ft from left end = 6700 and 1700 lb; 10 ft from left end = 1700 and -7300 lb; 17 ft from left end = -7300 lb and 3000 lb; at right end = 3000 lb. Bending moment: at left end = zero; 5 ft from left end = 33,500 ft-lb; 10 ft from left end = 42,000 ft-lb; 3 ft from right end = -9000 ft-lb. Inflection point = 4.23 ft from right end.

4-19-D $R_1 = 19,600$ lb, $R_2 = 8400$ lb. Shear at left end = zero; 8 ft from left end = -8000 lb and 11,600 lb; at right end = -8400 lb. V = zero at 8.4 ft from right end. Bending moment: at left end = zero; 8 ft from left end = $-32,000$ ft-lb; at 8.4 ft from right end = 35,200 ft-lb (maximum); at right end = zero. Inflection point is 16.8 ft from right end.

Chapter 5

5-4-A $c = 3.34$ in.

5-4-C $c = 4.13$ in.

5-9-A $I = 418.3$ in.^{4} $S = 45.8$ in.^{3}

5-9-C $I = 10.0$ in.^{4} $S = 3.0$ in.^{3} $r = 1.58$ in.

5-9-E $I = 431.4$ in.^{4} $S = 46.6$ in.^{3}

Chapter 6

6-2-A   Yes, $f_b = 22.1$ ksi

6-2-C   (a) 17.55 kips   (b) 9.21 kips

6-4-B   W 18 × 50

6-4-C   W 12 × 31 or W 14 × 30

Chapter 7

7-4-C   0.81 in.

7-4-E $(0.20 + 0.44) = 0.64$ in.

Chapter 8

8-3-B   Noncompact: $b_f/2t_f = 7.75 > 7.38$

8-4-A   (a) 80 kips (b) 51 kips (c) 34.5 kips (d) 55 kips

8-6-A   W 10 × 25

8-6-B   W 10 × 25

8-6-E   W 10 × 25

8-7-C   W 10 × 39

8-7-D   W 14 × 30

8-11-A   W 10 × 25

8-12-B   S 7 × 15.3; C 7 × 9.8; L 3½ × 3 × ⅜ or ⁵⁄₁₆

8-14-A   Bearing plate 8 × ¹³⁄₁₆ × 13

8-16-A   Yes, $f_b = 15.8$ ksi

8-16-C   No, actual deflection = approx. 0.5 in.

8-16-F   Deflection excessive, approx. 1.0 in.

8-16-I   W 8 × 20   (W 10 × 17 also, but not listed in Table 1-1)

8-16-L   Use bearing plate 9 × 1¹¹⁄₁₆ × 23

Chapter 10

10-3-C   117.2

10-6-A   435 kips

10-6-C   56.2 kips

10-9-D   TS 5 × 5 × ¼

10-10-A   78 kips

10-15-A   Plate 15 × 21 × 1⅜

10-16-D   W 12 × 58

10-16-E   96.7 kips

10-16-H   271 kips

10-16-J   Plate 14 × 17 × 1⅛

Chapter 11

11-6-C   11.4 kips

11-6-D   Double shear, 18.04 kips (Bearing = 18.6 kips)

Chapter 12

12-5-A $L_1 = 10.4$ in. $L_2 = 4.3$ in.

12-7-B   121 kips

Index

||||||||||||||||||||||||||||||||||||||||||||||||

Abbreviations, 5
Allowable deflection, 110
Allowable stress design, 107
Allowable stresses, 37
  for columns, 195
  for rivets and bolts, 228
  for structural steel, 42, 43
  in welds, 252
American Institute of Steel
  Construction, 1
American Society for Testing and
  Materials, 39
American Standard I-beams, 7, 9
  properties for designing, 18
Anchorage of trusses, 297
Anchors, wall, 163
Angles, structural, 10
  in tension, 238
  properties for designing, 22, 23
Arc welding, 250

Bars and plates, 12
Base metal, 252
Base plate, 213
Beam bearing plates, 159
Beam design procedures, 121, 127
Beam formula, 30, 101
Beam seats, 245, 263, 264
Beams, 45
  allowable moment chart, 151
  bearing plates, 159
  cantilever, 46, 81
  compact, 122
  continuous, 47
  definition of, 45

Beams (continued)
  design examples, 130–137
  design for bending, 105
  design procedure, 127
  fireproofing for, 176
  grillage, 217
  investigation of, 102
  lateral support for, 123
  laterally unsupported, 150
  loading, 47
  long spans and light loads, 152
  noncompact, 122
  overhanging, 46, 76
  restrained, 47
  safe load tables for, 140–145
  separators for, 159
  shear in, 126
  simple, 46
  types of, 45
  typical loadings, 82
  web crippling, 165
  wide flange, 7, 8
  with continuous action, 258
Bearing, enclosed, 227
Bearing plates, 159
Bearing-type connection, 225
Bearing strength of rivets, 226
Bending, theory of, 85
Bending factors, 210
Bending moment, 66
  diagrams, 68
  formulas, 83
Bending stresses, 29
Bolted connections, 221, 230
  beams to columns, 245

Bolted connections (*continued*)
 beams to girders, 239
Bolts, allowable stresses for, 228
 edge distance, 235
 high-strength, 232
 pitch of, 235
 ribbed, 231
 unfinished, 231
Bow's Notation, 289
Boxing, of welds, 254
Breaking strength, 34
Built-up sections, 95
Built-up separators, 159

Cantilever beams, 81
Carbon steels, 41
Center of gravity, 89
Center of moments, 50
Centroids, 89
Channels, 10
 properties for designing, 20
 safe load table for, 146, 148
Clip angle, 246
Column formulas, 193
Column shapes, 9, 190
Columns, 189
 allowable loads, 194
 base plates for, 213
 connections for, 245
 design of, 198
 eccentrically loaded, 209
 effective length, 192
 pipe, 199
 safe load table, 200
 splices for, 212
 structural tubing, 203
Compact sections, 122
Compression, 27
Compression members, 189, 289
Computations, accuracy of, 6
Concentrated load, 47, 70, 114
Connections, 242, 245, 247
 bearing-type, 225
 bolted, 230
 friction-type, 225, 232

Connections (*continued*)
 moment-resisting, 147
 riveted, 221
 seated, 245, 263, 264
 welded, 249
Continuous action, 258
Contraflexure, point of, 78
Coping and blocking, 239
Cover plates, 96
Crippling of webs, 165

Dead load, 172
Deflection, 109
 allowable, 110
 computation of, 111
Deflection coefficients, 115,
  118, 119
Deflection formulas, 83, 112,
  113, 117
Deformation, 32
Designations for structural
  shapes, 2, 13
Direct stress, 25
Distributed load, 47, 71
Double-angle struts, 205
 safe load table, 206, 207
Double shear, 226

Eccentric loading, 199
Eccentrically loaded columns, 209
 trial section for, 210
Edge distance, 235
Effective column length, 192
Effective net area of angles, 292
Elastic design, 299, 306
Elastic limit, 32
Electric arc welding, 250
Electrodes, 252
Enclosed bearing, 227
Equation of moments, 50
Equilibrium, laws of, 50
Equivalent loadings, axial, 210
 uniform, 147
 vertical, 280
Equivalent tabular loads, 83, 147
Extreme fiber stress, 30

Factor of safety, 36
Fasteners, 221, 242
Fiber stress, 29
Field welding, 256
Fillet welds, 251, 253
Fink Truss, 271, 283
Fireproofing for beams, 176
    weight of, 177
Fixed beam, 258
Flexure formula, 30, 87
Floor framing, 171
    design of typical, 177–182
Floors, loads on, 175
    weights of, 173
Flush top, 239
Framed beam connections,
    238, 242
Framing, floor, 171
Friction-type connection,
    225, 232

Gage lines, 234
Girder connections, 239
Girders, 45, 171
Grades of steel, 39
    for connectors, 44
Grillage beams, 217
Grillage foundations, 217

High-strength bolts, 232
High-strength steel, 39, 40
Hooke's Law, 34
Horizontal shear, 30

I-beams, 7, 9
Inflection point, 78
Investigation of beams, 102

Joists, open web steel, 182
    load table for, 184, 185

Lateral support of beams, 123
Laterally unsupported beams, 150
Laws of equilibrium, 50
Length, effective, 192
Light loads and long spans, 152

Lintels, 154
Live load, 174
Load factor, 311
Loads, 47
    computation of, 178
    concentrated, 47, 70, 114
    dead, 172
    distributed, 47
    equivalent tabular, 147
    equivalent vertical, 280
    equivalent wind and snow, 281
    live, 174
    minimum live, 175
    snow, 274
    superimposed, 106
    triangular, 83, 155
    truss, 272

Manual of Steel Construction, 1, 7
Masonry, strength of, 160
    weight of, 173
Materials, weights of, 173
Maximum bending moment, 73, 83
Maximum deflection, 83
Maximum shear, 83
Mechanical couple, 259
Mechanism, 309
Minimum live loads, 175
Miscellaneous shapes, 13
Modulus of elasticity, 34
Modulus, section, 92
Moment, 49
    bending, 66
    diagram, 68
    maximum, 73, 83
    negative, 69
    of a force, 49
    resisting, 85
    statical, 90
Moment arm, 50
Moment-connections, 247
Moment of inertia, 91
Movable partitions, 174
Multipass welds, 254

Negative bending moment, 69

Net section, 236
Neutral axis, 30
Neutral surface, 30
Nomenclature, AISC, 2–5
Noncompact section, 122

Open web steel joists, 182
    load table for, 184
Overhanging beams, 76

Panel point, 269, 270
Partitions, 173
    movable, 174
Pipe columns, 199
Pitch of rivets and bolts, 235
Plastic design, 107, 299, 306
Plastic hinge, 300, 303
Plastic moment, 300, 302
Plastic range, 299, 300
Plastic section modulus, 303
    computation of, 305
    table of, 307
Plate girders, 167
Plates and bars, 12
Plug welds, 261
Point of contraflexure, 78
Pratt truss, 271
Properties for designing, 14–23
    angles, 22, 23
    channels, 20, 21
    I-beams, 18, 19
    wide flange sections, 14–17
Properties of sections, 89, 93
    built-up, 95
    centroids, 89
    moment of inertia, 91
    radius of gyration, 95
    section modulus, 92
    unsymmetrical built-up, 98
Proportional limit, 34
Purlin, 269
    design of, 285

Radius of gyration, 95
Reactions, determination of, 53
Resisting moment, 85

Restrained beams, 309
Rigid frames, 315
Riveted connections, 221
    beams to columns, 245
    beams to girders, 239, 242
Riveting, 221
    symbols for, 223
Rivets, allowable stresses for, 228
    bearing in, 226
    conventional signs for, 223
    countersunk, 224
    edge distance, 235
    flattened, 224
    gage lines for, 234
    grades of, 44
    in trusses, 291
    pitch of, 235
    shear in, 226
    staggered, 235
    tension in, 228
Roof pitches and slopes, 270
Roof trusses, 269
    design of, 283
    parts of, 270
    types of, 271
Roofs, 173

S shapes, 9, 18
Safe loads, computation of, 102
Safe load tables, beams, 139
    channels, 146, 148
    W shapes, 140
    W and S shapes, 142–145
Sag rods, 286
Seated connections, 245, 263, 264
Section modulus, 92
    plastic, 303
Sections, structural, 7
Separators, 159
Shape factor, 306
Shapes, structural, 7
    designations for, 13
Shear, 27, 83
    computations for, 126
    diagrams, 59
    double, 226

Shear (*continued*)
  horizontal, 30
  in beams, 28, 30, 126
  in rivets, 226
  single, 225
  vertical, 57
Shear stress, 126
Shoe angles, 297
Shop welding, 256
Simple beam, 46
Single shear, 225
Slenderness ratio, 191
Slot weld, 261
Snow load, 274
Splices, column, 212
Staggered rivets, 235
Standard channels, 10
  properties for designing, 20
Standard I-beams, 7, 9
  properties for designing, 18
Statical moment, 90
Steel joists, 182
Steels, 39
  allowable stresses for, 42
  carbon, 41
  high-strength, 41
  properties of, 40
  structural, 39
Stiffeners, 165
Stiffness, 34
Strain, 32, 300
Strain hardening, 300
Strength of masonry, 160
Stress coefficients, 281
  table of, 282
Stress-deformation diagram,
    33, 330
Stress diagrams, 276
  notation, 289
  wind load, 279
Stresses, 27
  allowable, 37, 43
  bending, 29
  extreme fiber, 30
  in welds, 251
  kinds of, 27

Stress-strain diagram, 33, 300
Structural shapes, 7
Structural steels, 39, 40
  carbon, 41
  corrosion-resistant, 42
  high-strength, 41
Structural tees, 11
Struts, 189, 205
Superimposed load, 106
Symbols for welds, 266

Tees, 11
Tension, 25, 27
Tension members, 238
Threaded fasteners, 229
Triangular loading, 83, 155
Trusses, 269
  compression members in, 289
  end bearing and anchorage, 297
  equivalent vertical loadings, 280
  identification of joints and
      members, 289
  loads on, 272
  notation for, 289
  purlin design, 285
  rivets in, 291
  roof, 269
  spacing of, 272
  stress coefficients for, 281, 282
  stress diagrams for, 276
  tension members in, 291
  types of, 271
  weights of, 273
  welded joints in, 294
Type 1 construction, 247
Type 2 construction, 247

Ultimate strength, 32
  design theory, 299
Unfinished bolts, 231
Unit stresses, 25

Vertical shear, 57

W shapes, 8, 14–17
Wall anchors, 163

Walls, 173
Warren truss, 271
Web crippling, 165
Weights, of materials, 173
  of trusses, 273
Weld reinforcement, 251
Welded connections, 249, 264
Welded joints, 250
  design of, 255
Welded truss connections, 294
Welds, 251
  fillet, 251, 253

Welds (*continued*)
  multipass, 254
  plug and slot, 261
  stresses in, 251
  symbols for, 265, 266
Wide flange shapes, 7, 8
  properties for designing, 14
Wind load, on roofs, 275
Wind load stress diagram, 279

Yield point, 32, 34, 300

Decimals of an Inch for Each 64th of an Inch, with Millimeter Equivalents

Fraction	64ths	Decimal	Millimeters (approx.)	Fraction	64ths	Decimal	Millimeters (approx.)
....	1	0.015625	0.397	....	33	0.515625	13.097
1/32	2	0.03125	0.794	17/32	34	0.53125	13.494
....	3	0.046875	1.191	....	35	0.546875	13.891
1/16	4	0.0625	1.588	9/16	36	0.5625	14.288
....	5	0.078125	1.984	....	37	0.578125	14.684
3/32	6	0.09375	2.381	19/32	38	0.59375	15.081
....	7	0.109375	2.778	....	39	0.609375	15.478
1/8	8	0.125	3.175	5/8	40	0.625	15.875
....	9	0.140625	3.572	....	41	0.640625	16.272
5/32	10	0.15625	3.969	21/32	42	0.65625	16.669
....	11	0.171875	4.366	....	43	0.671875	17.066
3/16	12	0.1875	4.763	11/16	44	0.6875	17.463
....	13	0.203125	5.159	....	45	0.703125	17.859
7/32	14	0.21875	5.556	23/32	46	0.71875	18.256
....	15	0.234375	5.953	....	47	0.734375	18.653
1/4	16	0.250	6.350	3/4	48	0.750	19.050
....	17	0.265625	6.747	....	49	0.765625	19.447
9/32	18	0.28125	7.144	25/32	50	0.78125	19.844
....	19	0.296875	7.541	....	51	0.796875	20.241
5/16	20	0.3125	7.938	13/16	52	0.8125	20.638
....	21	0.328125	8.334	....	53	0.828125	21.034
11/32	22	0.34375	8.731	27/32	54	0.84375	21.431
...	23	0.359375	9.128	....	55	0.859375	21.828
3/8	24	0.375	9.525	7/8	56	0.875	22.225
....	25	0.390625	9.922	....	57	0.890625	22.622
13/32	26	0.40625	10.319	29/32	58	0.90625	23.019
....	27	0.421875	10.716	....	59	0.921875	23.416
7/16	28	0.4375	11.113	15/16	60	0.9375	23.813
....	29	0.453125	11.509	....	61	0.953125	24.209
15/32	30	0.46875	11.906	31/32	62	0.96875	24.606
....	31	0.484375	12.303	....	63	0.984375	25.003
1/2	32	0.500	12.700	1	64	1.000	25.400